SURVIVING
EXTREME
WEATHER

SURVIVING EXTREME WEATHER

THE COMPLETE CLIMATE CHANGE PREPAREDNESS MANUAL

MYKEL HAWKE
AND JIM N. R. DALE

Skyhorse Publishing

Skyhorse Publishing books may be purchased in bulk at special discounts for sales promotion, corporate gifts, fund-raising, or educational purposes. Special editions can also be created to specifications. For details, contact the Special Sales Department, Skyhorse Publishing, 307 West 36th Street, 11th Floor, New York, NY 10018 or info@skyhorsepublishing.com.

Skyhorse® and Skyhorse Publishing® are registered trademarks of Skyhorse Publishing, Inc.®, a Delaware corporation.

Visit our website at www.skyhorsepublishing.com.

10 9 8 7 6 5 4 3 2 1

Library of Congress Cataloging-in-Publication Data is available on file.

Cover design by David Ter-Avanesyan
Cover photo credit:
    Flood: Photo by Saikiran Kesari on Unsplash
    Fire: Photo by Caleb Cook on Unsplash
    Dust: Photo by Marcus Kauffman on Unsplash
    Satellite: Photo via Shutterstock
Interior photo credit:
    Section 1 Title: Photo by Michale Held on Unsplash
    Chapter 1 Situation: Photo via iStock
    Chapter 1 Survival: Photo by David Law on Unsplash
    Science of the Sun: Photo by Fabian Struwe on Unsplash
    Chapter 2 Situation: Photo via iStock
    Chapter 2 Survival: Photo by Malachi Brooks on Unsplash
    Section 2 Title: Photo by Matt Gross on Unsplash
    Chapter 3 Situation: Photo via iStock
    Chapter 3 Survival: Photo by Curioso Photography on Unsplash
    Chapter 4 Situation: Photo via iStock
    Chapter 4 Survival: Photo by Kei on Unsplash
    Section 3 Title: Photo by Todd Trapani on Unsplash
    Chapter 5 Situation: Photo by Lucy Chain on Unsplash
    Chapter 5 Survival: Photo by Max Larochelle on Unsplash
    Science of Water: Photo by Clay Leconey on Unsplash
    Chapter 6 Situation: Photo via iStock
    Chapter 6 Survival: Photo by Wade Austin on Unsplash
    Section 4 Title: Photo by NASA on Unsplash
    Chapter 7 Situation: Photo by NASA on Unsplash
    Chapter 7 Survival: Photo via iStock
    Chapter 8 Situation: Photo via iStock
    Chapter 8 Survival: Photo by Nikolas Noonan on Unsplash
    Chapter 9 Situation: Photo by Peter Schulz on Unsplash
    Chapter 9 Survival: Photo by Mattieu Jannon on Unsplash
    Section 5 Title: Photo by Nadia Genzhyi on Unsplash
    Chapter 10 Situation: Photo via iStock
    Chapter 10 Survival: Photo by Thijs Stoop on Unsplash
    Chapter 11 Situation: Photo via iStock
    Chapter 11 Survival: Photo by James Wainscoat

ISBN: 978-1-5107-7798-9
Ebook ISBN: 978-1-5107-8017-0

Printed in the United States of America

Authors' Note

We hope this book will serve our fellow mankind as we face unprecedented global climate change. Whether you believe in it being a natural or man-made occurrence, the facts amply show we are experiencing extreme weather with catastrophic impacts . . . and no matter where you are, we all share one thing in common—we all want to survive.

From Mykel:
I would like to say thanks to the team for making this happen; thanks to my wife and son for their love and support; and thanks to you, who had the vision to see and read this for readiness. Let's hope the leaders now and next generation to come can make the change.

We'd also like to give a special credit and appreciation to Ms. Kim Martin for her invaluable contributions.

From Jim:
Many thanks to all those with their "fingerprints" over this manual; you know who you are, and nothing at all would have happened without you!

Table of Contents

Here Comes the Sun

The world has changed.
I see it in the water.
I feel it in the Earth.
I smell it in the air.
Much that once was is lost,
For none now live who remember it.

A huge shout out to J. R. R. Tolkien and director Peter Jackson of *Lord of the Rings* fame for those masterful words that provide a fitting beginning to the film *The Fellowship of the Ring* . . . and also our book![1]

Professor Tolkien wasn't of course writing about modern-day climate change on this beautiful multicolored Earth of ours, he was instead referring to a time in his mythical Middle Earth when, after a respite of several hundred peaceful years, the forces of evil were once again rising against the good and well-meaning inhabitants of his mystical world.

Professor Tolkien had visions of a torrid world ravaged and laid bare by war and blind destruction (that is if the evil forces were to have sway). His dark visions in the real world were galvanized by time spent in the mud-laden, stinking trenches of the Somme,

---

1 Jackson, Peter. 2001. *The Lord of the Rings: The Fellowship of the Ring.* United States: New Line Cinema. Quote adapted from *The Lord of the Rings: Return of the King* by J. R. R. Tolkien.

France during World War 1. The destruction and decay that lay all about him could easily be compared to the desolation upon the passing of a tornado, a hurricane, a major flood, or a wildfire. They are all equal in their deep ugliness and the toll they may bring.

So, we can, in a fair way, draw a link between Tolkien's multi-layered world and the perilous threats that befell them to our own world and the potentially hazardous position we find ourselves facing today. In the main, it is due to the shortsightedness, selfish intent, and failures of mankind to heed the global warming warnings issued by climate scientists and bodies such as the World Meteorological Organization and NASA. Just as the good people were at war with the dark forces that occupied parts of Middle Earth, so are we in our own way, and it is a war that, in the long-term, the well-meaning people of today's Earth may not be able to win!

Of course, we are not up against a Dark Lord, orcs, goblins, or other noire creatures of the abyss (let's perhaps wait and see what materializes out of the debris of globally warmed Earth, though). For us, a much more powerful enemy has arisen, an enemy that has, since the dawn of mankind, lurked in one guise or another, but has now been strengthened and emboldened by our own recklessness and disregard for the consequences. Our new enemy is the vengeful and dangerous dark side of Mother Nature. She is adorned in clothing of many colors, and she has numerous weapons of engagement. She moves around our globe at will, scouring and defacing anything that lies in her path. She is ultra-powerful and gaining vigor with every passing year. Entire ecosystems are at the mercy of her will, and humans and all other life forms can be no match for her wrath, as and when she cares to show it. She is what we have created and now, if we want to be safe, if we wish to survive and prosper, we must become wise to her. We must learn and act appropriately, we must be informed and alert, able to run when we need to and defend ourselves when we have to.

And so we arrive at the meaning of this book. It is appropriate to point out that what follows is not about climate change or global warming per se. The arguments there will continue to rage between the scientific community and those willing to listen and act, versus the self-interest of the fossil fuel merchants, their supporters, and the downright ignorant—perhaps we should call them the modern-day orcs?

As far as our guidance is concerned, we are assuming that the Titanic hit the iceberg long ago, and both the ship and the iceberg are sinking at an undetermined rate. Yes, some governments and local communities can move around the deckchairs and provide life jackets to whomever they can, while others in more exposed and perilous parts won't be so fortunate in the skirmishes, battles, and wars to come. The purpose of our guidance, and irrespective of your climate change views, is to help you and your loved ones to be safer and survive whatever the dark side of Mother Nature might throw at you, wherever in the world you may find yourself.

Before we plow on too far and sound like we are giving the last rights on a path to Armageddon, here's some very basic physics to justify why we are concerned to help where we can and provide guidance where necessary. Understand that, as the land, the oceans, and the ice continue to warm and melt at the rate they have over the most recent thirty years or so, molecules in their various states move more quickly when that extra heat is applied, in turn allowing for greater energy be released. That greater energy has led to a steady but ongoing dislocation of known climate zones (rapid warming of the Arctic and Antarctic are two indisputable examples of this), in turn creating two pivotal outcomes:

1. Bigger weather
2. Chaos

The *bigger weather* presents itself in more powerful hurricanes and typhoons, fiercer tornadoes, longer-lasting droughts, rampaging wildfires, devastating floods, more numerous and increasingly severe thunderstorms, sudden and smothering snow and ice accretion events, unexpected and more numerous avalanches, debilitating and expansive sand and dust storms, unyielding and choking pollution.

We have already borne witness to many such events and there are more to come.

The *chaos,* meanwhile, translates into undetermined, often unforeseen events, stretching beyond pure weather itself, such as wildlife- or human-inspired impacts, new, unfathomable diseases and plagues, country- or continent-wide food and water shortages, enforced human and wildlife migrations, unanticipated skirmishes, wars, and other undesirables of a grossly horrific nature. All of the above, and what can be deemed "normal" weather-related perils of one kind or another, may also join forces such as extreme heat and wildfires, and at these times you may have to play extra safe or battle on more than one front.

Safety and survival are the objectives here—*your* safety and survival and that of your family and friends. So, this book is for you and whoever else within your compass may care to read it, particularly if you live in, or are planning to visit, a known or developing hazard-prone country or region.

Finally, please listen up to this carefully, as it underscores what is to follow within our book at all times: Safety and survival can *never* be guaranteed. Nobody—no book and no expert—can ever prepare you or your loved ones for every kind of situation, as many weather or related situations will be sudden, chaotic, and of an idiosyncratic nature. It's at these times that composure, common sense, and confidence in yourself and those around you make these merits the best weapons of self-defense. However, by reading and digesting what might be appropriate for you to follow, we sincerely hope that if there is a time of weather-related distress and danger in your life, we may be of some assistance in getting you out the other side, preferably in one piece.

Thank you for placing your trust in us. May your god and/or your science go with you . . . not forgetting your good fortune. We all need a little bit of that!

Best regards,
Mykel and Jim

SECTION I
HEAT AND FIRE

You are your own walking talking thermometer, but the very best thermometers are attached to a thermostat.

1

Hotter than Hell

HEAT—THE SITUATION

Of all the elements of the weather entwined within a warming world, extreme heat is arguably the most obvious phenomena that we might face.

Of course, it's fair to say that the Earth itself has undergone many major changes and deviations in terms of its temperature profile throughout its 4.6 billion years of existence. The causes were many, and all were through different forms of natural evolution or sudden catastrophic events. Moreover, it's perfectly clear that mankind's sticky fingers were not responsible for any of those numerous changes . . . until now.

Historically speaking, the main reasons for Earth's major surface temperature changes are four-fold and all what we can fairly term as being *natural events*.

- The Sun's intensity, or flux. Over many millennia, meaningful changes have occurred in terms of the Sun's energy output which have, in turn, warmed or cooled the atmosphere, the landmasses, and the oceans here on Earth. However, this is not the case for mankind's relatively short-term existence, though some would argue that other fluxes, such as solar winds and sunspot activity, have resulted in some short-term temperature fluctuations. The Sun *is* getting hotter or more luminous as it heads to its own inevitable oblivion in 5–6 billion years' time, but these changes have been, and will be, absolutely minuscule in our own lifetimes and are not part of the shifting temperature debate.

- The shape of the Earth's orbit around the Sun, combined with the ever-changing angle of its tilt, has resulted in periodic global warming and global cooling events. That's over many thousands of years, of course, and absolutely not over the past few decades.

  The changing configuration of our continents and oceans, Gondwanaland, Pangaea, Tethys, Laurasia, and all those stages of evolution. Planet Earth is a dynamic shape-shifter, and those shifts have resulted in wide-scale climate aberration, all of them firmly in the dim and distant past.

• Catastrophic, mostly short-term events, such as asteroid impacts and volcanic eruptions. For example, the relatively recent eruption of Mount Pinatubo in 1991 cooled the troposphere in the northern hemisphere by as much as 0.7 C for around two years. And then there was the extinction of the dinosaurs. They finally kicked the bucket after surviving for some 165 million years, due (and it is widely accepted) to climate change brought about by a devastating combination of a huge asteroid impact and major volcanic eruptions.

On the Earth's cosmic calendar, (credit the great Carl Sagan), modern-day mankind arrived on the scene as we are today, a few seconds to midnight on December 31, with primitive man only first emerging around midday. Which means, that if we talk about the survival of mankind, then compared to the dinosaurs' existence of five long days (165 million years), we are but a breath of wind within a long-lived tempest.

Why is all that important? Putting it simply, we have a fifth, very recent reason for accelerating temperatures: the burning of carbon-based fossil fuels that are continuing to ramp up global temperatures at an exponential rate, far beyond what we might term normal. That is perilous, and not only a threat to our own existence on Planet Earth, but to every living organism, which is why it is crucial.

To offer a simple way of understanding why a warming world is perilous to all that live upon it, here is something to think about. The Earth has many living and breathing ecosystems, all existing within a certain physical/chemical profile with acutely defined boundaries, including humans. Our bodies are programmed to happily exist at a predetermined internal body temperature range. What if the heat outside meant that it was difficult to keep to those critical internal levels during a prolonged severe heat wave, for example? Even a short-lived, two-degree Celsius rise in your internal temperature would likely put you in bed, if not in hospital, while a three-degree Celsius rise would no doubt put you in your grave. The Earth continues to warm and will very soon pass the 1.5 degrees Celsius average rise since the industrial revolution kicked in on the way toward more disturbing numbers that will then see whole ecosystems collapse. That's simply because the tolerance ranges of a host of living things will have been breached and may in the end include our own. That is why the relatively modest aims of the Intergovernmental Panel on Climate Change (IPCC) to limit the average global temperature rise to below a couple of degrees is so very critical, and sadly, is almost certain to be missed. Those small heat numbers aren't small; they are immense beyond comprehension.

That's the big background, so let's discuss the ramifications of the big heat of our todays and tomorrows for when it comes knocking at your door. We humans live and prosper within a critical internal corridor between 97°F (36.1°C) and 99°F (37.2°C). Meander only a little from that relatively small range, and our bodies easily become negatively impacted, eventually to the point of death.

When should we be concerned about heat? When meteorologists mention air temperatures, we are mainly talking shade and not full sunshine. So, if shade temperatures surpass 95°F (35°C), we need to have concern. If and when they breach 100°F (37.8°C), then we are well past the point of our own internal temperature, and it will make itself known. Add in high humidity and/or exposure to full sunshine, and we enter into a potential nightmare zone.

Of course, one man's meat is another man's poison; we inhabit and prosper in different parts of the globe, with widely varying temperatures and equally varying tolerance levels. You should be aware and irrespective of the cause of excessive temperatures on any given day, what you and those around you are able to safely tolerate, noting that the old, the young, and people with underlying health problems are likely to suffer more quickly and in various physiological and psychological ways as the thermometer rises.

Finally, my job is to give you the rhymes, reasons, and the broad direction of travel. Mykel's job, on the other hand, is to expertly advise you and those around you on how to stay safe and, in the worst-case scenarios, survive. I will, therefore, not venture into Mykel's world within my advice, save for this one solitary occasion. The name of the game when battling excessively high temperatures is to keep your cool and stay there, ideally within those ideal body temperature parameters. My past employers, the British Royal Navy, and their scientists, were responsible for this particular tip, which is now utilized by a number of professional sports people as they begin to feel the heat. When you feel you are genuinely overheating, dip your fingertips (or, for that matter, your toes) into a bowl of icy water for three to four minutes. Those sensitive and very responsive digits will gradually and painlessly transfer a degree or two of your internal body heat into the icy water, helping you to revert back into your comfort zone. Now over to Mykel, for his super-heated take.

HEAT—THE SURVIVAL

Jim has made a good case, well researched and based on strong scientific facts. Now that we have some info on what extreme heat looks like and can do, we will take a look at some things you can do to help mitigate risks, manage heat-related disasters, and prepare for surviving when things heat up.

SCIENCE, FACTS, AND HISTORY

For this chapter and others, we're going to cover it from all angles—left, right flanks, rear and frontal, from above and below—so we can give you as many advantages as possible.

The key to surviving the increasing occurrences of drastic changes in climate and extreme weather events is to understand them from a practical perspective and then plan and execute tactical responses.

Categorized by elements—fire, water, wind, and earth—events are explained with recommendations anyone can follow to better deal with the impacts and repercussions and, in doing so, minimize the risks of damage to life and property. Included are lists of items every person, home or business, and vehicle should always stock; first aid kits, afflictions, and applications; structural and landscape precautions and improvements; philosophies to help family, friends, and neighbors get through any environmental ordeal.

I'll share personal stories, experiences, and anecdotes that always bring some bearing for consideration, as well as provide some comic relief for all the other doom and gloom that this subject matter can bring, so we'll apply some levity to counter the gravitas. And we'll use a standard, military-type checklist approach for fast reference throughout the book and in appendixes.

First, we have to take a macro view of things to get our head wrapped around the full situation in its entirety. There are many subcategories of heat-related disasters, and each warrants its own book on the matter, but we're just going to do a one-over-the-world sort of approach here. Most of these things are not new in any way. However, the scope, scale, frequency, and impacts are truly new.

For one, we just crossed the threshold of 8 billion people on the planet. Let's hover on this concept for a moment as it will drive all further conversations about disasters if, for no other reason, than there are simply more people who will be affected by climate change, extreme weather, and natural disasters *because* there are more people.

I've been teaching survival for three decades now, and I have always said that population was going to be the major issue in our lifetime. When I was born in 1965, there were only 3 billion humans on the planet. If I live a normal lifespan, there will be 9 billion. Think about that for a minute. What took all of humanity's existence up to 1965 to reach 3 billion will now have tripled in one human lifetime. When I pass, that will likely double, or more, in the next human's lifetime.

When more people are in harm's way, it takes more people to respond, more resources, and more time. So, a normal cleanup operation might take months, or years, and with more frequency, then there ends up being a never-ending cycle of disaster and recovery. This often leads to fatigue on the part of the rescuers, strain on the part of the authorities, and burnout of the population, decreasing sensitivity and impacting overall support.

This doesn't apply only to climate-related disasters. Maybe more concerning is the growing population's effect on resources. Consider an analogy I'll use to make this complex topic very simple. If the world is one hill, and on that hill is one fruit tree, and for a while that fruit tree can feed everyone until there are too many mouths to feed, then we enter into businesses and corporations that try to take advantage of those resources and the systems for their own gain. When that gets out of control, people begin to protest, and that ultimately leads to conflicts and wars. These can be coups, rebellions, revolutions, or outright, full-scale war. These are sometimes settled by victories, not always by the good guys or by intervention, like NATO, the UN, or some world powers working together or unilaterally.

The reason for covering all that is simple: once you understand all conflict in the context of a struggle for resources, what I am about to say will make a lot more sense.

I have been in many wars—drug wars, gold wars, diamond wars, etc.—but in the past, they could all be boiled down to oil. Oil is a precious resource for the entire planet, at least for now. Electric, solar, and many other forms of power are being developed, but for the moment, it is still primarily combustion engines that have the power to move large amounts of cargo on trains, planes, trucks, and ships.

The more we use oil, the less there is in reserve. But that's not the bad news. I believe for all the new problems we face with each generation, more new ideas and solutions we will always find. I have faith in humanity. We will keep finding new ways to move large amounts of goods and learn to rely less and less on oil. So, that's actually okay. The real problem is not oil . . . it's water.

About 2.5 percent of the Earth's water is freshwater, but only a small fraction of that is readily available as drinkable water. According to the United States Geological Survey (USGS), approximately 0.5 percent of the Earth's total water supply is available as freshwater that can be accessed by humans. This freshwater can be found in rivers, lakes, and groundwater sources, and only a small portion of it is easily accessible and of high enough quality for drinking. Therefore, while the Earth has a vast amount of water, only a small percentage of it is suitable for human consumption without treatment or filtration.

A rapidly growing population will need much more water than ever before and, between waste and pollution, we're actually decreasing our vital, life-giving supplies. Again, many brilliant minds are working to find solutions, but the fact remains for now, we can't create a water alternative.

Now, back to that heat problem. If we're talking forest fires, because of extreme heat and drought, we need more water to fight them. They're happening more and more, burning longer and harder, destroying more, which ends up making us have less water, not only from the water used to put the fires out, but from the lack of fresh water being produced due to more of the environment being destroyed. That's a very vicious downward spiral. Then you look at the damage to the atmosphere allowing more of the sun's deadly rays to penetrate and continue causing even more damage. Then, the droughts caused by that means less water for crops and herds, which means less food for more people. Then, you look at the damage to human skin causing cancers, the hotter temps meaning more needed for air conditioning, which consumes more energy, causing rises in costs and greater demands for water.

So, fires, heat, drought, loss of crops, starvation of people and herds are all bad news, to be sure. As an intelligence specialist, I know you can't find solutions until you review all the problems. Now, we have done that. Let's look at a few numbers so you have some parameters.

Here are some recent examples of heat disasters. In the modern era, several locations have experienced extreme heat waves, resulting in some of the highest temperatures recorded for humans. Here are examples:

- Furnace Creek Ranch, California, USA (July 10, 1913): Furnace Creek Ranch in Death Valley holds the record for the highest temperature ever recorded in North America, at 134°F (56.7°C).
- Tirat Tsvi, Israel (June 21, 1942): Tirat Tsvi recorded a temperature of 129°F (54°C), which is the highest temperature ever recorded in Asia.
- Kebili, Tunisia (July 7, 1931): Kebili recorded a temperature of 131°F (55°C), which is the highest temperature ever recorded in Africa.
- Mitribah, Kuwait (July 21, 2016): Mitribah recorded a temperature of 129.2°F (54°C), which is the highest temperature ever recorded in Asia.

The hottest places on Earth on a yearly average are typically located in deserts, where the combination of high temperatures and low humidity can create extremely hot and arid conditions. The top three hottest places on Earth on a yearly average are:

1. Death Valley, California, USA: Death Valley holds the record for the highest temperature ever recorded on Earth, at 134°F (56.7°C). The average temperature in Death Valley

during the summer months is around 115°F (46°C), and the annual average temperature is around 77°F (25°C).

2. Dasht-e Lut, Iran: Dasht-e Lut is a large salt desert in southeastern Iran that holds the record for the highest surface temperature ever recorded on Earth, at 159.3°F (70.7°C). The annual average temperature in Dasht-e Lut is around 113°F (45°C).

3. Dallol, Ethiopia: Dallol is a hydrothermal field in northern Ethiopia that experiences some of the hottest and most inhospitable conditions on Earth. The annual average temperature in Dallol is around 95°F (35°C), with temperatures regularly exceeding 120°F (49°C).

Other hot places on Earth include the Sahara Desert, the Arabian Peninsula, and the Australian Outback, where temperatures can regularly exceed 100°F (38°C) for months at a time.

The highest survivable temperature for a human is around 60°C (140°F). As you can see, we're already approaching that in some places. At temperatures above this range, the body can overheat, leading to heat exhaustion or heat stroke. These conditions can cause symptoms such as nausea, dizziness, and confusion, and can be life-threatening if not treated promptly.

It is important to note that these temperature ranges are for naked humans without any external sources of heat or cooling. With proper clothing and equipment, humans can survive in more extreme temperatures for shorter periods of time.

It is difficult to determine the exact highest-temperature fever a human has had and survived, as it can vary depending on individual factors such as age, overall health, and the underlying cause of the fever.

In general, a fever is considered to be a body temperature above 100.4°F (38°C). Fevers can be caused by a variety of factors, such as infections, autoimmune diseases, or reactions to medications.

While rare, some people can experience very high fevers that can be life-threatening. Fevers above 105°F (40.6°C) can cause damage to the brain and other organs, and can be particularly dangerous for young children, the elderly, and people with weakened immune systems. It is difficult to give a precise temperature at which humans die from fever, as it can vary depending on individual factors, such as age, overall health, and the underlying cause of the fever.

It is key to know that, at 102°–104°, you should make treatment an emergency, strip all clothing, get ice and water and fans or AC on them, and get help. At 106°, people are about to die. Do whatever it takes to cool them down.

However, it is important to note that most fevers can be treated with medication and supportive care, and rarely reach such high temperatures. It is always recommended to seek medical attention if a fever is accompanied by severe symptoms or lasts for an extended period of time.

Now, what to do about all this? In special forces, we are world-renowned for being deep planners. We always go at least five layers deep when we plan. We make a Primary and Alternate, a Contingency and Emergency (PACE), and finally, a GTH, or Go to Hell plan. I highly recommend that you, the reader, take this same time-tested and battle-proven methodology. It is beyond the scope of this book to teach that kind of planning, but use the five-point contingency planning for all you do and you will have the very best chances of surviving anything that comes your way.

That said, for the most part, two simple plans can usually always be developed on a moment's notice, and it's how we handle things that get thrown at us in the middle of any mission. We always look first at what the worst-case scenario could be, assess our resources and make a plan. Then, we bring it down a notch to what is the most likely scenario, and we plan our COAs, or Courses of Action, accordingly. And we can apply those here.

The worst-case scenario for heat disasters is the Sun or a comet burns us up with fires and volcanoes. In this case, the plan is simple, make your peace but fight your hardest. Never give up hope. Some may survive.

The most likely scenario is more of the same, annually repeating, ever-increasing in severity and duration, impacting us more and more. That's not great news, but there are things we can do. Let's take a look at some.

LOCATION

The obvious answer is that deserts will become harsher. Water will be more scarce, more expensive, and more regulated. I personally believe nations will go to war more over water than they ever have over oil. Let that sink in. I also believe that more of the Arctic is going to melt and make more green land available there. Of course, that would mean more water in the oceans, so higher water levels means many low-level tropical places will become flooded and under water. So, relocation is one major option to consider. However, if that is not possible, let's look at potential scientific solutions that can help you as an individual.

SCIENCE AND SOLUTIONS

When it comes to solving problems, there are always human beings out there working on it. Some professionally hired by corporations with hopes of profits, some hobbyists tinkering, some accidental discoveries, some academics who simply want to solve a challenging problem. So, there is always hope, and things are always changing.

The point of that last sentence is important; by the time this book comes out, there will be new solutions for old problems. And it is a continuous process, so one must continuously keep an eye out for things that may help you and your family.

For example, if you live in a fire-prone area, it seems prudent to make sure you have fire preventions. Not just fire extinguishers, perhaps specialized water sprinklers, fireproof paints and materials, in the home, on the home, around the home. You can make fire breaks in your wood line if forested, have early warning systems, set up notices in your newsfeeds, make sure you have a good evacuation plan, not just the route and destination, but how you travel—what you bring, what you leave, how you leave it, fireproof safes, fire shelters in the home, under the home, outside the home, bunkers with water, oxygen, extreme heat prevention, and cool air or ventilation control.

It would not be possible in one book to provide all the solutions for everyone's unique and individual situations. Rather, our purpose is to give you info, ideas, directions, guidance, and hope. But you have to care, give it the time, put in the homework, or not. Maybe you simply can't afford a lot of fancy options. Maybe you don't have time to do a great deal of deep planning. But I would strongly recommend, if you love your family and home, please do what you can to be as preventive, prepared, responsive and ready should some catastrophe come calling.

To get the creative problem-solving juices flowing, I will include some old and some new ideas for each problem set we cover; some will actually apply to many events, situations, and scenarios. If they work for you, check them out and, if not, maybe they will inspire you to do your own research and diligence to develop a survival strategy for your specific needs.

TACTICAL

So first, the number one issue to cover—that we really shouldn't need to—is don't stare at the sun! Staring at the sun can cause damage to your eyes. The sun emits harmful ultraviolet (UV) rays that can cause serious and sometimes irreversible damage to the retina, the delicate layer of tissue at the back of the eye responsible for vision.

Looking directly at the sun, even for a few seconds, even during a partial solar eclipse or when using binoculars or other viewing devices, can cause a condition known as solar retinopathy, which can result in blurry vision, distortion, and even permanent damage to the retina. Special eclipse glasses or solar filters that are designed to block harmful UV rays are the only safe way to view the sun during an eclipse or other astronomical event.

If you do accidentally stare at the sun, it is important to seek medical attention immediately if you experience any symptoms such as vision loss or distortion.

More broadly, to protect yourself from the dangerous effects of the sun and heat, there are several measures you can take:

1. Wear protective clothing: Wear lightweight, loose-fitting clothing that covers as much skin as possible, such as long-sleeved shirts, pants, and hats.
2. Use sunscreen: Apply a broad-spectrum sunscreen with at least SPF 30 to all exposed skin, including your face, neck, and ears. Reapply every two hours, or more often if swimming or sweating.
3. Seek shade: Stay in the shade as much as possible, especially during peak sun hours between 10 a.m. and 4 p.m.
4. Stay hydrated: Drink plenty of water and other fluids to stay hydrated and replace fluids lost through sweating.
5. Avoid alcohol and caffeine: Both alcohol and caffeine can dehydrate you, so it is best to avoid them during periods of high heat.
6. Use fans or air conditioning: Use fans or air conditioning to help cool your home or workspace and reduce the risk of heat exhaustion or heat stroke.
7. Take breaks: Take frequent breaks and rest in a cool, shaded area if you must be outdoors during high heat.
8. Be aware of symptoms: Be aware of the symptoms of heat exhaustion and heat stroke, such as headache, dizziness, nausea, and confusion. Seek medical attention immediately if you experience these symptoms.

By taking these protective measures, you can help reduce your risk of heat-related illnesses and enjoy the summer safely.

REMEDIES

If you are experiencing overheating, there are several things you can do to cool yourself down in an emergency:

1. Get to a cooler environment: Move to an air-conditioned room, or at least to a shady area if you are outside.

2. Remove excess clothing: Strip down to lightweight, breathable clothing, or remove clothing altogether, if possible.
3. Hydrate: Drink cool water or other non-alcoholic, non-caffeinated beverages to replenish fluids lost through sweating.
4. Use a fan: If available, use a fan or other source of cool air to aid in evaporative cooling.
5. Apply cool water: Wet a towel with cool water and apply it to your skin, or take a cool shower or bath to lower your body temperature.
6. Seek medical attention: If your overheating is severe or you experience symptoms such as confusion, dizziness, or rapid heartbeat, seek medical attention immediately.

It is important to note that overheating can be a serious condition and may require medical attention in some cases. Prevention is key, so it is important to stay hydrated and avoid excessive exposure to high temperatures whenever possible.

It is important to seek medical attention if symptoms do not improve, or if you experience severe symptoms, such as blistering, swelling, fever, confusion, seizures, difficulty breathing, or if the burn covers a large area of your body. In addition, it is important to take steps to prevent heat injuries, such as staying hydrated, wearing protective clothing, and avoiding prolonged exposure to high temperatures.

HOME

If you are experiencing extreme heat in your home, there are several emergency actions you can take to reduce the temperature and prevent heat-related illnesses:

1. Turn on fans: Use fans to circulate the air in your home and create a breeze.
2. Close curtains and blinds: Closing curtains and blinds can help block out sunlight and reduce heat gain from windows.
3. Open windows: If the outdoor temperature is cooler than

the indoor temperature, open windows to allow cooler air to enter the home.

4. Use air conditioning: If you have air conditioning, turn it on and set it to a comfortable temperature.

5. Stay hydrated: Drink plenty of cool water or other non-alcoholic, non-caffeinated beverages to prevent dehydration.

6. Take a cool shower or bath: Taking a cool shower or bath can help lower your body temperature and provide relief from the heat.

7. Use ice packs or wet towels: Apply ice packs or wet towels to your neck, armpits, and groin to help cool your body down.

8. Seek cooler shelter: If your home is still too hot, consider going to a public place with air conditioning, such as a shopping mall, library, or community center.

It is important to note that extreme heat can be a serious condition and may require medical attention in some cases. Prevention is key, so it is important to take steps to stay cool and hydrated during hot weather.

There are several new technologies available that can help make your home cooler and more comfortable during hot weather:

1. Smart thermostats: Smart thermostats allow you to control the temperature of your home remotely using a smartphone app or voice commands. They can also learn your preferences and adjust the temperature accordingly, helping to save energy and keep you comfortable.

2. Solar-powered air conditioning: Solar-powered air conditioning systems use solar panels to generate electricity, which powers the air-conditioning unit. This can help reduce energy costs and make your home more sustainable.

3. Radiant barriers: Radiant barriers are reflective materials that can be installed in your attic to reflect heat away from your home. This can help reduce heat gain and keep your home cooler.

4. Smart windows: Smart windows use technology that can automatically adjust the amount of light and heat that enters your home, based on the time of day and temperature. They can also be controlled remotely using a smartphone app.

5. Evaporative cooling: Evaporative cooling systems use water to cool the air in your home, similar to how sweating cools the human body. They can be more energy-efficient than traditional air-conditioning systems and are especially effective in dry climates.

6. Heat pumps: Heat pumps can both heat and cool your home using a single unit. They work by extracting heat from the air outside your home and transferring it inside during the winter, and reversing the process during the summer to cool your home.

These are just a few examples of the newest technologies available to make your home cooler. It is always important to consider factors such as energy efficiency, cost, and maintenance requirements when choosing the best cooling system for your home. But there are some ways to turn the sun and heat into positives for your home. The sun and heat can be harnessed to benefit your home in several ways. Here are some tips for taking advantage of the sun and heat:

1. Use natural light: Use natural light to reduce the need for artificial lighting during the day. Install skylights, or use light-colored window treatments to allow more natural light into your home.

2. Install solar panels: Install solar panels to generate electricity and reduce your reliance on fossil fuels. Solar panels can also help reduce your energy bills and increase the value of your home.

3. Use passive solar heating: Use passive solar heating to warm your home in the winter. Install south-facing windows to allow sunlight to enter your home and warm the interior.

Use thermal mass, such as concrete or tile flooring, to absorb and store heat.

4. Install a solar water heater: Install a solar water heater to heat your water using the sun's energy. Solar water heaters are cost-effective and can reduce your energy bills.

5. Install a green roof: Install a green roof, which is a layer of vegetation installed over a waterproof membrane, to insulate your home and reduce energy costs.

6. Install shading devices: Install shading devices, such as awnings, shutters, or shade sails, to block the sun's rays and reduce heat gain in your home.

By taking advantage of the sun and heat, you can reduce your energy bills, increase the value of your home, and reduce your carbon footprint.

VEHICLE

If you must plan on traveling a long distance by foot or car in extreme heat, it can be challenging and potentially dangerous. Here are some tips to help you stay safe and comfortable during your journey. Most of the same precautions for protection and prevention of overheating apply here. In addition:

1. Take breaks: Take frequent breaks to rest and cool down. Find a shaded area or an air-conditioned building to rest in, if possible.

2. Plan ahead: Plan your route in advance, and check weather conditions before leaving. Avoid traveling during the hottest part of the day, if possible.

3. Monitor your health: Monitor yourself and others for signs of heat exhaustion or heat stroke, such as dizziness, nausea, or confusion. Seek medical attention immediately if symptoms occur.

4. Keep emergency supplies: Keep emergency supplies on your person or in your vehicle, such as extra water, sunscreen,

first aid kit, and a fully charged cell phone, with you at all times.

Now, if you're caught out in the worst of it, such as being stranded in the high desert during an extreme heat wave, in addition to aforementioned precautions, the following steps can help you stay safe:

1. Find shade: Seek shade immediately to avoid exposure to the Sun's rays. Look for natural shade, such as a rock overhang or a tree, or create your own shade with a tarp or other materials.
2. Stay low: Stay close to the ground to avoid exposure to the Sun's rays. The air temperature can be significantly cooler near the ground.
3. Signal for help: Use a mirror or other reflective object to signal for help. Create a signal fire if possible, but be aware of the risk of starting a wildfire.
4. Preserve energy: Avoid unnecessary physical activity and conserve your energy. Rest during the hottest part of the day and avoid traveling during this time.
5. Stay positive: Stay calm and positive. Mental attitude can have a significant impact on physical endurance and survival.

Here are a few examples of the newest technologies available to make your car cooler. It is always important to consider factors such as cost and compatibility with your vehicle when choosing the best cooling technology for your car.

1. Solar-powered car fans: Solar-powered car fans can be placed on the dashboard or in the windows of your car and use solar energy to power a fan that circulates the air inside the car.
2. Ventilated seats: Ventilated seats use fans or air conditioning to blow cool air through the seats, keeping you cool and comfortable while driving.

3. Smart air conditioning: Some car manufacturers now offer smart air-conditioning systems that can adjust the temperature based on the outside temperature and the number of passengers in the car.
4. Window tinting: Window tinting can help block out sunlight and reduce heat gain inside the car. Newer technologies, such as ceramic window tinting, can be more effective at blocking heat than traditional window tinting films.
5. Heat-reflective paints: Heat-reflective paints can be applied to the exterior of the car to reflect sunlight and reduce heat absorption, keeping the car cooler.
6. Remote start: Many newer cars now offer remote-start functionality, which allows you to start the engine and turn on the air-conditioning system before getting into the car.

If you are experiencing extreme heat in your car, there are several emergency actions you can take to reduce the temperature and prevent heat-related illnesses:

1. Open windows and doors: Open all windows and doors to allow the hot air to escape and fresh air to circulate.
2. Turn on air conditioning: If your car has air conditioning, turn it on and set it to a cool temperature.
3. Use sun shades: Use sun shades on the windshield and windows to block out sunlight and reduce heat gain.
4. Park in the shade: If possible, park your car in a shaded area to reduce the amount of direct sunlight it receives.
5. Use a wet towel: Wet a towel with cool water and place it on your neck or forehead to help cool your body down.
6. Use a fan: If you have a portable fan, use it to circulate the air and create a breeze.
7. Stay hydrated: Drink plenty of cool water or other non-alcoholic, non-caffeinated beverages to prevent dehydration.
8. Avoid leaving children or pets in the car: *Never* leave children

or pets unattended in a parked car, as the temperature can rise quickly and lead to heatstroke or death.

It is important to note that extreme heat in a car can be a serious condition and may require medical attention in some cases. Prevention is key, so it is important to take steps to stay cool and hydrated during hot weather and avoid prolonged exposure to high temperatures.

And the last bit on tech for staying cool is stuff that can help you personally. There are several new technologies available that can help you stay cooler in extreme heat:

1. Cooling vests: Cooling vests contain special cooling packs or gel inserts that can be frozen and then worn to help regulate your body temperature and keep you cool.

2. Portable air conditioners: Portable air conditioners can be used indoors or outdoors and can help lower the ambient temperature in a specific area. Some models are small enough to be used in a car or on a camping trip.

3. Personal fans: Personal fans can be worn around your neck or clipped onto your clothing to provide a steady stream of cool air.

4. Cooling towels: Cooling towels are made of special materials that can be soaked in water and then wrung out to create a cool, wet cloth that can be draped over your neck or forehead.

5. Battery-powered misting fans: Battery-powered misting fans can be used indoors or outdoors and spray a fine mist of water to help lower the ambient temperature and keep you cool.

6. Smart clothing: Some companies are developing smart clothing that uses built-in sensors and cooling technology to help regulate your body temperature and keep you comfortable in extreme heat.

These are just a few examples of the newest technologies available to help you stay cool in extreme heat. It is always important to consider factors such as cost, effectiveness, and portability when choosing the best cooling technology for your needs.

MORE ABOUT THE SCIENCE OF THE SUN

The Sun is key to our existence in that most plants require it for photosynthesis to grow and feed us with their food and their production of oxygen. So, the Sun's impact of potentially drying up our precious, life-giving water makes it important for us at this juncture to review some basic science and facts relating to the Sun.

SUN FACTS

Everyone knows about the Sun. But there are some important scientific things to understand how it can impact you, me, us, and the Earth, and what it can't do. Most folks know about solar flares. When the Sun has a massive eruption on its surface, it can send out waves that can temporarily impact our atmosphere. These solar flares can send out electromagnetic radiated pulses that can impact satellites and cause things like the GPS to be off by a few yards, too. But that's about it.

The Sun also has its equivalent to our volcanos, called a CME or Coronal Mass Ejection. These solar explosions propel bursts of particles and electromagnetic fluctuations into Earth's atmosphere. Those fluctuations could induce electric fluctuations at ground level that could blow out transformers in power grids. A CME's particles can also collide with crucial electronics onboard a satellite and disrupt its systems. But the good news is that it can't reach our planet in the form of heat.

We do have government agencies and many other academic entities and organizations that study these and, much like with hurricanes, we can't always predict when, where, and how they'll strike, but we usually have enough notice to take measures.

When these solar flares and coronal mass ejections happen, there are a few types of issues that result. Some of the sunburst energy is in the form of ionizing radiation, some of the energy is magnetic energy, and some is ultraviolet radiation. The Earth's atmosphere does a pretty good job of protecting us from much of the potential sun damage. But it would appear we are destroying that to a degree, which is impacting us in the form of more cancer, more drought, and more polar ice caps melting, causing floods and changes in climate forcing all life to adapt, or not.

CMEs can create proton storms. Protons are the positively charged particles from the nucleus of an atom. The explosion that creates the CME accelerates the protons around the Sun to nearly the speed of light. These protons carry dangerous amounts of energy that can break chemical bonds.

The charged particles from a proton storm interact with the atmosphere and cause spectacular changes to the atmosphere, known as the aurora borealis or northern lights. The light and X-rays from CMEs reach Earth's atmosphere in about eight minutes. Usually, the Earth's atmosphere protects people from proton storms.

Proton storms can interfere with ham radio communication and damage satellites, causing short circuits in electrical systems and shutting down computers. Because powerful proton storms have the potential to cause a lot of damage to technology that we rely on, the National Aeronautics and Space Administration (NASA) tracks the Sun's activity so we can prepare for large proton storms. NASA also researches new ways to protect astronauts in space, such as those on the International Space Station (ISS).

Now, for a bit more geeking out on my part to give you a better understanding of solar impacts on human survival.

There are five classes of solar flares, according to National Oceanic and Atmospheric Administration (NOAA). Their designation depends on the intensity of X-rays emitted. Each class letter represents a ten-fold increase in energy output, similar to the Richter scale that measures the strength of earthquakes.

NASA says the X-class flares are the most powerful solar flares. Then there are M-class flares that are ten times smaller than X-class flares, then C-class, B-class, and finally A-class flares, which are too weak to significantly affect our planet Earth.

Within each letter class, there is another scale from one to nine, which gives the assessment greater precision, with larger numbers representing more powerful flares within the class.

However, X-class flares can break this nine-point rating mold with higher ratings since there is no class more powerful than X-class. According to NASA, a 2003 solar flare was so powerful, it overloaded the sensors measuring it. The sensors reported an X28 are before cutting out.

Fortunately, X-class flares occur, on average, about ten times per year, and flares as powerful as the one recorded in 2003 are even less likely. And more good news is that the Sun has a natural cycle of a major spike in activity near these levels every eleven years. So, everyone over the age of eleven has already survived one of these at least.

A bit more information, as this is a big subject; after all, early man worshipped the Sun as it was seen to be all-important to all life.

Different types of flares, particularly X-class flares, affect Earth, satellites, and even astronauts. But, luckily for us, A- and B-class solar flares are the most common and are also the weakest of the solar are classes, too feeble to affect Earth in any significant way. C-flares are also fairly weak, having little or no effects.

With the two largest classes of flares, things start to get a little more interesting.

Strong M-class and X-class flares can trigger coronal mass ejections—a large release of plasma and magnetic eld from the sun.

This can disrupt our magnetosphere and result in geomagnetic storms. Such geomagnetic storms can lead to auroras closer to the equator than is possible during calm conditions.

In 1989, according to NASA, a large solar are accompanied a coronal mass ejection and hit Earth, plunging the entire province of Quebec, Canada, into an electrical blackout that lasted twelve hours. The solar eruption triggered a geomagnetic storm on Earth, resulting in aurora borealis that could be seen as far south as Florida and Cuba. Now that is impressive.

I am also a former special forces communication sergeant, well-schooled in HF radios, antenna theory, and how it all impacts the radio waves. Radio blackouts are caused by higher electron densities in the lower ionosphere of our atmosphere. That's the layer through which radio waves pass. The enhanced electron density causes radio waves to lose more energy as they travel through the layer, preventing them from reaching higher layers that refract the radio signals back down to Earth.

Such radio blackouts are the most common space weather event to affect Earth, with approximately two thousand minor events happening each solar cycle.

The severity of radio blackouts depends on the strength of solar flare and is ranked R1 to R5 on the NOAA Solar Radiation Storm Scale. One represents a minor event and five is an extreme event. R5 events can lead to radio blackouts on the entire sunlit side of Earth and can last several hours. Luckily, for us, we experience, on average, less than one R5 per solar cycle. Now, we need to take a quick look at documented historical extremes to help put things into perspective and, hopefully, provide some level of peace once it is all better understood.

NASA released some photos of sun spots in 2023. Some of them were as large as six planet Earths! They pose no threat to us; think of them more like a temporary cool spot (relatively speaking), and that impacts normal solar flare activity. I'm also an FCC licensed radio operator, and I won't go into a deep-dive on phenomenon like Sporadic E, but it's neat to read up on if you're curious.

For those worried about a worst-case scenario, we do have one of the biggest recorded solar flares in history, which happened on September 1, 1859, and caused an aurora that was seen around the world—but we survived. We had another big one in 1921 that was ten times bigger than the bad one of 1989, which took out a Canadian province for twelve hours. They say if that happened today, it would affect 130 million folks and take four to twelve months to recover at 1 to 2 trillion dollars, so the threat is real, just not fatal.

They key takeaway in all this, is that the Sun is vital to life but can cause death, albeit not likely to ever be direct; it comes mostly in the form of cancer from long-term exposure without protections, occasional electronic damage that could cause fatal accidents, and the indirect, long-term impacts of climate change due to human-induced damage to our atmospheric protection layers.

HUMAN SUN FACTS AND FACTORS

Firstly, we all know about sunburns. I think everyone has had at least one negative experience with this, and that's good for my purposes as a teacher. We can all see and agree that the sun can physically impact us. And, fortunately, there's lots we can do to protect ourselves.

The use of hats and sunglasses are standard. But hats should have at least three inches of brim around to give the best protection. Sunglasses are rated by their UV protection, a light spectrum we can't see but can feel; they are not rated based on darkness. Spend a bit here and get real UV glasses. We also use sun lotion. That is its own study—how it affects skin, the water we get into, how to reapply, and all this is important to know, but not our task here. And then, there are clothes. Many clothes these days also have UV or SPF sun-protection factor. Clothes are better protection than lotion.

The strength of the Sun's UV rays reaching the ground depends on a number of factors, such as:

- Time of day: UV rays are strongest in the middle of the day, between 10 a.m. and 4 p.m.

- Season of the year: UV rays are stronger during spring and summer months. This is less of a factor near the equator.
- Distance from the equator (latitude): UV exposure goes down as you get farther from the equator.
- Altitude: More UV rays reach the ground at higher elevations.
- Cloud cover: The effect of clouds can vary, but it's important to know that UV rays can get through to the ground, even on a cloudy day.
- Reflection off surfaces: UV rays can bounce off surfaces like water, sand, snow, or pavement, leading to an increase in UV exposure.

An easy way to remember this, best for teaching:

Slip on a shirt. Slop on sunscreen. Slap on a hat. Wrap on sunglasses.

Exposure to the Sun produces vitamin D, which has many health benefits. It might even help lower the risk for some cancers. Your skin makes vitamin D naturally when you are in the sun. How much vitamin D you make depends on many things, including how old you are, how dark your skin is, and how strong the sunlight is where you live. At this time, doctors aren't sure what the optimal level of vitamin D is. A lot of research is being done in this area. Whenever possible, it's better to get vitamin D from your diet or vitamin supplements rather than from sun exposure, because dietary sources and vitamin supplements do not increase skin cancer risk and are typically more reliable ways to get the amount you need. All this talk about UV means we need to take a bit of a deeper dive into what that really is, as it does matter to our survival-planning strategies.

Ultraviolet radiation is energy from the sun that reaches the earth as visible, infrared, and ultraviolet (UV) rays. Ultraviolet A (UVA) is made up of wavelengths 320 to 400 nm (nanometers) in length. Ultraviolet B (UVB) wavelengths are 280 to 320 nm. Ultraviolet C (UVC) wavelengths are 100 to 280 nm. Those numbers are only included to help understand a bit of how they are different.

But only UVA and UVB ultraviolet rays reach the earth's surface. The earth's atmosphere absorbs UVC wavelengths, for now.

UVB rays cause a much greater risk of skin cancer than UVA. But UVA rays cause aging, wrinkling, and loss of elasticity. UVA also increases the damaging effects of UVB, including skin cancer and cataracts.

The way it works is that ultraviolet rays react with melanin. This is the first defense against the sun. That's because melanin absorbs the dangerous UV rays that can do some serious skin damage. A sunburn develops when the amount of UV damage exceeds the protection that the skin's melanin can provide.

A suntan represents the skin's response to injury from the Sun. A small amount of sun exposure is healthy and pleasurable. But too much can be dangerous. Measures should be taken to prevent overexposure to sunlight. These preventive measures can reduce the risks of cancers, premature aging of the skin, the development of cataracts, and other harmful effects. This is especially true as we diminish our atmosphere and its protective properties.

TACTICAL

For Sun, you want to look at protections for your home (where you may have to stay for static survival) and for your car (in case you have to go into a mobile survival situation).

These were addressed under Heat, but also, you need to have water, be near water, have a plan to procure water, and a plan to process it and make it safe.

In addition, there are a few actions we can take to mitigate some aspects of the potential damage from solar flares. These can destroy some electronics, so this is a place we can also take positive action; even though it could seem a bit extreme at first, it's actually a very practical approach and, with the threat of modern nuclear war and the associated EMP, these measures give some small indemnity for both.

Solar flares are not uncommon. They are usually weak and do not affect the earth. A massive solar flare can do a lot of damage, especially to the power grid, since we don't have any protection from a massive solar flare. The likelihood of a massive solar flare happening in our lifetime is small.

The human body is not affected by solar flares, so you do not need to directly protect yourself, but you do need to protect electronic devices, as they can be destroyed. Here's a technique for making a poor-man's Faraday cage, or protection for electronics from EMP. Unplug your electronic devices, place them in a cardboard box, and wrap the box around with aluminum foil. Be sure to place the box on the ground; this way, the electric charge from the solar flares can safely be discharged into the ground.

You can keep your more valuable electronic devices safe from a solar flare by using a Faraday duffel bag, certified for EMPs. Just remember, protecting your phones and computers won't do you too much good if the grid and communications themselves are knocked out.

But the very best way to protect your devices is with a Faraday cage. This is an enclosure formed by a conductive material that blocks external static and non-static electronic fields by channeling electricity along and around, but not through it, while providing a constant voltage on all sides of the enclosure. Since the difference in voltage is the measure of electrical potential, no current flows through the interior of it, thus protecting your devices.

With all that said, as covered in Heat, it is important to protect your vehicle from solar flares, using those preventions and safety measures. Remember, prolonged exposure to the sun and solar flares can be harmful to your vehicle, causing paint to fade and interiors to crack and discolor.

STRATEGIC

This section is where you do the biggest picture planning for your scale of operations. For example, if you decide to buy a completely mechanical hand-crank window type car, or to move to a new state, or even a new country, seeking shelter from the sun or for the resource of water. This will be very different for every person, but it should be given some planning consideration for your long-term survival strategies. With all that, let's set a common ground for preparation and readiness against extreme sun action.

To protect your home from the sun and solar flares, here are some best things you can always do:

1. Install window films: Window films can help reduce the amount of heat and UV radiation that enters your home through windows. Look for films that are designed specifically to block UV rays.
2. Use curtains or blinds: Curtains or blinds block direct sunlight and reduce heat gain in your home.
3. Install awnings or shades: Awnings or shades on the exterior of your home provide shade and reduce heat gain.
4. Plant trees: Plant trees strategically around your home to provide shade and reduce heat gain. Make sure to choose trees that are appropriate for your climate and won't cause damage to your home.
5. Use reflective roofing materials: Use roofing materials that are highly reflective to reduce heat gain in your home. This can help reduce cooling costs and extend the life of your roof.
6. Solar shields: Solar shields or blankets can be used to protect your home from solar flares. These shields are made of highly reflective materials that can deflect harmful radiation.
7. Regular maintenance: Regular maintenance of your home, including cleaning, painting, and sealing, can help protect it from the damaging effects of the sun.

Remember, prolonged exposure to the sun and solar flares can be harmful to your home, causing fading, cracking, and other damage. Taking these precautions can help protect your home from these damaging effects and keep it in good condition for years to come.

MY STORIES

For this portion, I have a few personal anecdotes I'd like to share.

When I was a communications (commo) man in special forces, we had to deal with some phenomena called Sporadic E. Sporadic E (or E Skip) occurs when patches of ionization form in the E region of the ionosphere, which can reflect radio waves back to Earth. This can result in radio signals traveling much farther than they would normally, allowing for long-distance communications that would otherwise be impossible.

Sporadic E is particularly useful for amateur radio operators, who can take advantage of these unpredictable propagation conditions to make contacts over much greater distances than would normally be possible. However, it can also cause interference with other radio communications, particularly during solar maximum when the ionosphere is more active.

One period of time when we were in Texas practicing making commo shots or communications around the country, we encountered this Sporadic E phenomena and, at the same time, we encountered another phenomenon entirely—our food was disappearing.

Since this was not tactical training, we didn't have to post security on the military installation in the middle of nowhere. So, we slept and awoke to find our meals were gone. They were MREs (military food), and we had plenty more, so no worries. We figured maybe it was taken by a coyote or something.

We broke camp, hiked miles away, set up camp, and did our communications shots. When done, we went to sleep and, again, awoke to food thievery. But, this time, they ripped open all the meals, and only ate the good stuff! We couldn't eat the leftovers, as they were gnawed upon! So, repeat, break camp, march, make camp, commo, eat, and sleep, but this time, we pulled guard duty.

I was on duty when I found the culprit—one of the largest raccoons I ever saw. He looked like a small bear. I didn't want to shoot him, so I used the slingshot we carried to get our antennas high into the air by shooting fishing weights tied to a fishing reel attached to the wrist rocket. I grabbed a rock, shot it with the slingshot right in the hind quarters where I aimed. The raccoon didn't even stop eating or look at me! I got a bigger rock, hit the same spot, then he slowly put down the food and looked at me. I respected that, so I went to bed, didn't wake the next guy, and let him eat what he wanted; he deserved it. The next day, we went very far away, and he didn't follow us again.

FAMILY

Here are some good considerations for both basic and high-tech ways to protect your family from a massive solar flare.

A solar flare is a brief eruption of intense high-energy radiation from the Sun's surface. In extreme cases, it can cause geomagnetic storms that can disrupt communication systems, satellites, and power grids on Earth. To protect your family from the potential effects of a catastrophic solar flare, consider both basic and high-tech approaches:

BASIC APPROACHES:

1. Emergency preparedness: Have a well-stocked emergency kit that includes food, water, medical supplies, and other essentials to sustain your family for at least seventy-two hours. Include flashlights, batteries, and a battery-operated or solar-powered radio to stay informed during power outages.

2. Communication plan: Establish a family communication plan to stay connected during an emergency. Identify a meeting place and an out-of-town contact person.

3. Physical protection: Install surge protectors for your home's electrical system to prevent damage from power surges caused by geomagnetic storms.

4. Backup power: Invest in a generator or solar panels with battery storage to provide backup power during prolonged power outages.

HIGH-TECH APPROACHES:

1. Early warning system: Subscribe to space weather alerts and monitoring services to receive real-time updates on solar activity, so you can take precautionary measures in advance.

2. Faraday cage: Store critical electronic devices, such as smartphones, laptops, and radios, in a Faraday cage or a makeshift metal container. This can shield them from the harmful effects of electromagnetic radiation generated by a solar flare.

3. Backup communication systems: Use satellite phones or radio communication systems, such as ham radios, to communicate in case of disruptions to traditional communication networks.

4. Data protection: Regularly back up your important digital data on an external hard drive or cloud storage, and keep a copy in a safe and shielded location.

5. Educate yourself and your family: Learn about solar flares, geomagnetic storms, and their potential impacts on Earth's systems. This knowledge will help you make informed decisions on how to protect your family during such events.

Remember, preparedness is key. The more proactive measures you take, the better equipped your family will be to handle the effects of a catastrophic solar flare.

FRIENDS AND NEIGHBORS

Here are some of the best basic and high-tech ways to protect yourself, your family, and your immediate neighbors, as well as support your friends' networks from a catastrophic solar flare. Helping your neighbors prepare for, and cope with, the potential effects of a catastrophic solar flare is essential for fostering a resilient community. Note that many of these practices, in general, are applicable to other events and situations.

BASIC APPROACHES:

1. Share information: Educate your neighbors about the potential risks associated with solar flares and geomagnetic storms. Share reliable resources and websites where they can learn more and stay updated on space weather alerts.

2. Community emergency planning: Collaborate with your neighbors to develop a community emergency plan, including designated meeting points, communication strategies, and resource sharing agreements.

3. Assist with emergency supplies: Offer assistance to neighbors in assembling their own emergency kits. You can also organize a community effort to stockpile essential supplies, like food, water, and medical items, in a central location for shared access.

4. Community drills: Organize neighborhood drills to practice responding to potential power outages and other impacts of a solar flare. This will help everyone become familiar with their roles and responsibilities during an emergency.

HIGH-TECH APPROACHES:

1. Communication network: Establish a neighborhood communication network using alternative communication methods, such as walkie-talkies or ham radios, to stay connected during potential disruptions to traditional communication systems.

2. Group purchasing: Organize a group purchase of critical items like surge protectors, backup power supplies, or Faraday cages to benefit from bulk discounts and ensure everyone has access to these protective measures.

3. Technical assistance: Offer to help neighbors protect their electronic devices and data by providing guidance on using Faraday cages, backing up data, and implementing surge protection measures in their homes.

4. Local monitoring and alerts: Set up a local space weather monitoring system, or subscribe to alert services as a

community. Share real-time updates and warnings with your neighbors through a designated communication channel.

5. Building a resilient community is a collective effort. By working together and sharing resources, knowledge, and support, you can help protect not only your family but also your neighbors from the potential impacts of a catastrophic solar flare.

VEHICLE

The best way to protect your car against solar flares is with conductive fabric; these come in different forms and are specifically designed to stop RF energy. These conductive fabrics can also protect your motorcycle, generators, and any other electrical devices which are not hooked into the power line directly.

The conductive fabrics typically can provide up to 40–50 dB of shielding, which is very good. Note that a small tear renders them unusable in the case of a solar flare. Keep in mind that some of these meshes have silver and nickel in them for those who are allergic. Solar Flare EMPs can affect most modern cars. So, if you have a fully mechanical backup, that's just one less worry.

Remember, while these technologies can provide some protection from solar flares, they should not be relied upon as the only means of protection. It's important to stay informed about solar activity and to follow any advisories or warnings issued by local authorities. In addition, it's always a good idea to have backup plans in place in case of power outages or other disruptions caused by solar flares.

I just want to leave you here with a simple thought—the Earth is heavily reliant upon the Sun to exist as it does currently, hosting all life as we know it, for now. The Webb Telescope is changing everything by establishing realities of physics that are off the charts of anything we ever imagined.

I take great heart in this. The reason is simple, whether you believe in a god or not, if our earth can be as diversified as it is and we've learned so much already, about microbes to deep sea, to insects and a vastness of all things fungi, etc., then imagine how much more is out there in universes and galaxies that we can't even imagine?

I choose to believe that God is real, and in a real form our science and beings can't yet understand . . . that God is good, that it is a universal spirit of goodness, an architect of intelligent design, so that even the programming of leaves to evolve, or how a human can learn enough to understand microscopes to telescopes as a scale applied to all known endeavors, and our specialization within every manner of thing will grow exponentially, each somehow connecting to everything else. As such, we continue to move forward toward the goal of fulfilling our potential . . . if we survive. And, that's where I'll leave this. If our Sun explodes, our fate is sealed. If our Sun perishes slowly, very likely, so will we . . . but, we may find a way, despite it, or, because of it.

2

Beds Are Burning

FIRE—THE SITUATION

There's a song, "Firestarter," by a British band called The Prodigy and, although not necessarily about a fire starter in the true sense of the meaning, the robust piece of artistry—and particularly the way it is delivered by the band—aptly fits around the incredible and terrifying subject we are about to tackle.

I cannot begin to downplay this hideous phenomenon in any way, given its capacity to ravage the land and everything in it and on it, including all life forms. For anyone reading this who has been caught up in such a misadventure, you will know exactly what I mean. Smoke smothers, fire consumes, and both kill!

So, how does it all begin, and what meteorological factors might be involved? Well, in order to set the scene let's borrow the name from another band, this time the 1970s American outfit Earth, Wind, and Fire, as a nice and simple way to recall the major elements.

When we look into the causes and effects of wildfires, there are several components, and the first of these is the status of the Earth itself or, if you prefer, the ground surface and anything that lies or grows upon it. That could be trees, branches, twigs, leaves, grass, shrubs, crops, moss, peat, litter, debris, animal habitats, and homes. The list is large, but these can all be encompassed in a single word, and that is *fuel*. The status of these varied fuels is vital, because as we all know, damp, or particularly wet surfaces do not easily combust, nor does fire easily spread across damp or wet areas. Therefore, dry and low humidity conditions, and more especially, parched or baked surfaces are necessary factors, as is the abundance of the fuel, such as forests, woods, heath land, or even relatively closely packed combustible buildings. In September 1666, The Great Fire of London was a perfect example of a ravenous wildfire, consuming more than thirteen thousand wooden houses, all with pitched straw roofs and all interconnected.

Note the month of that terrible event, September. It followed what was a particularly hot and dry summer, so it was a tinder-dry environment, despite the close proximity of the River Thames. The majority of wildfires tend to occur in late summer or very early autumn in both hemispheres when the Earth, and all that is on it, is at its driest. But, and as a fair warning to you, that's not in every case, so please be aware at other times of year also when parched conditions may prevail.

The most at-risk regions of the world to wildfires tend to be in, or adjacent to, densely forested, wooded, or bush-laden regions, with Russia/Siberia, Australia, California, Brazil, Indonesia, and Canada historically most prone. However, even within these regions, climates vary, so there are few climatic exceptions beyond the tallest mountains and sand or ice deserts. Moreover, as climate change takes hold, wildfires are advancing into places you might not think are at risk, including Alaska and several coastal destinations around the Mediterranean Sea. The bottom line here is that wildfires are becoming more common and spreading in their geographical reach. As we write this book, the globally over-heated summer of 2023 bears absolute testament to this "new abnormal" with devastating life-taking wildfires spread far and wide across countries and regions including Canada, Hawaii, Algeria, Portugal, the Canary Islands, and the Greek Islands of Rhodes and Corfu. There may have been more after the ink was set and there will be more in the years ahead, and as with some of the cases mentioned above, they will move into villages, towns, and cities; where fire fuel is plentiful.

We now move to the reasons for a wildfire to occur, some meteorological and some not. There are at least two eternal ones that come with weather wrapped all around them. The first of these is a lightning strike. Hotter than the surface of the Sun and hotter than lava on Earth, lightning strikes to trees are perhaps the most common catalyst, but other tall structures such as buildings, telegraph poles, and the ridges of hills also provide a compliant recipient.

The second natural cause has a quieter guise and is far more surreptitious, and that is combustion by direct or indirect heating by the Sun's rays. There are many organic items that can spontaneously combust due to excessive insolation, dead leaves being one and lying straw in fields being another. I could go on to mention coal seam, lava, or meteor strike fires, these being three other natural causes, but all of those and any others you might care to mention are uncommon.

What is increasingly commonplace are wildfires caused by human influence; yes, our sordid paws are by far the largest factor. It is said that in modern times, over 80 percent of wildfires are man-made. That's an astounding figure, but it's also a very sad one. Arson is a big part of that, and negligence also figures highly, with campfires, discarded cigarettes, and agricultural burning three of the main instigators.

So, we now have a fire and enough dried fuel to burn the kingdom to come. But there is one other factor that truly makes a wildfire wild and that's wind! Wind is the transporter, and the higher the wind speed, the faster and farther the wildfire is likely to spread. Add in higher velocity winds, and we have a recipe for unpredictability, chaos, and acute danger. The tragic August 2023 wildfires on the island of Maui were largely caused by such high-strength winds, as a hurricane passed hundreds of miles to the south of this Pacific retreat. The 50–60 mph sustained winds drove what were previously higher-level rural fires, from east to west across the island, rapidly blow-torching the census-designated place of Lahaina, with many poor souls unable to escape.

The end game may arrive in many forms, with death and absolute destruction at the top of the bad tree. If human efforts fail to eradicate or contain a wildfire, then we have to look to nature to do it for us in the form of dousing rainfall or those destructive winds turning about and pushing the head of the fire back where the fuel has already been exhausted. But in our own case, that hasn't yet happened, and we are now going back before any flame has taken hold, because safety and survival strategies begin well in advance of any spark.

FIRE—THE SURVIVAL

In the military, we generally have three layers of structure that help us manage the complexities of warfare:

> **Strategic**: What the overall campaign commander focuses on.
> **Operational**: What regional leadership focuses on.
> **Tactical**: What the troops on the ground focus on.

These have been battle-tested over millennia, and serve as a good start point for how you can approach your survival planning as you are waging war against the cold—because you will be fighting for your life.

Let's look at some strategic approaches, which will be the big-picture plans you make, like protecting your home and family. Operational strategies, such as movements or responses to events, and then the tactics, which are those personal things you can do when facing the threat in the moment.

TACTICAL

We are going to focus on what you can do for your people, pets, homes, and gardens and/or yards to protect it from fire. This is where you look at everything within your control to give your loved ones the best chances of surviving.

If you are camping in an area affected by a wildfire, it is important to pack essential items to protect yourself:

1. A portable radio: A battery-powered or hand-cranked radio can provide you with important updates on the wildfire and evacuation orders.
2. A go bag: Prepare a bag with essential items, including important documents, medications, a change of clothes, and personal items.
3. An N95 mask: Use an N95 mask to protect your lungs from smoke and ash.
4. Fire-resistant clothing: Wear long pants, long sleeves, and boots made of fire-resistant materials.
5. Nonperishable food and water: Pack enough food and water to last for at least three days.
6. First aid kit: Include supplies for treating burns, cuts, and other injuries.
7. Emergency shelter: A lightweight, waterproof tent or tarp can provide shelter from smoke and ash.
8. Fire extinguisher: Bring a small fire extinguisher to help put out small fires.
9. Map and compass: In case of evacuation, it's important to know the location of nearby roads, trails, and other landmarks.
10. Headlamp or flashlight: A reliable light source can help you navigate in smoky conditions.

Remember to always follow the instructions of local authorities and evacuate if necessary. Be aware of your surroundings and stay safe. It's also important to keep as many of the same items in your vehicle as possible.

OPERATIONAL

This one takes a bit wider view, usually the affected area, be it the block, neighborhood, town, valley, city, etc. In short, these are all the things that could impact you and your plans. Can the town respond? Is there a fire department nearby, are they active or volunteer, what is their past performance? You'll need to make this assessment each time for each category. For example, you may have a rock-solid EMS system, but a part-time, faraway fire department; both factors impact your planning for both what you can do, what you can expect or not, and which ones you'll need to have a more solid set of backup plans for.

Remember, the best way to protect yourself from a wildfire is to stay informed, follow evacuation orders, and have a plan in place. Stay safe and be prepared.

STRATEGIC

This is always the big picture. Ask yourself: *If this event got bigger than what my local authorities could handle, what kind of response can I expect?* FEMA usually comes to mind, but also your state governor may mobilize the National Guard, or your local mayor may ask for help from other regional municipalities. It is *always* best to plan on *no help.* Keep in mind what they could do, or may do, and have it as a factor, but I strongly encourage everyone to be completely self-sustaining and autonomously able to do their own rescue, whenever possible.

First, let's look at what are good indicators a fire is coming. There are several indicators, including:

1. Smoke: Smoke is the most obvious indicator that a fire is nearby. If you see smoke in the distance, it's important to pay attention and be prepared.
2. Flames: If you can see flames, a fire is already in your area. This is a clear sign that you should evacuate immediately.
3. Heat: A sudden increase in temperature, or a feeling of warmth on your skin can indicate that a fire is nearby.

4. Strong winds: Strong winds can quickly spread a fire, making it more difficult to contain. If you notice strong winds, be aware that a fire could spread quickly.

5. Falling ash: If you see ash falling from the sky, it could be a sign that a fire is nearby.

6. Animals behaving unusually: Animals may sense a fire before humans do and may start behaving unusually or run away.

7. Unusual silence: If you live in a wooded area, and the birds and animals suddenly go quiet, it could be a sign that a fire is approaching.

Remember, if you notice any of these indicators, it's important to act quickly and follow evacuation orders. Always listen to local authorities and be prepared to evacuate at a moment's notice.

The reality is, most of us can't afford to do and be our own FD, PD and/or EMS, much less a search and rescue or a security quick response force, so we must rely on the authorities. That's part of why we pay taxes, and it's up to you to hold your elected officials accountable for how they spend and plan for emergencies. Get involved, stay involved; it's what made America great and it's the only thing that makes America work. So, your planning doesn't end at your property line, it extends from you all the way to the president and everything in between. Being a good citizen means being responsible, participative, and caring.

MY STORIES

Firstly, a very direct fire-impact story isn't anything like losing a home or life, but when I first moved to California, there were bad fires in San Diego, and for days I was shovelling soot and ash off my car like people up north shovel snow.

Next, I want to share a small bit of trivia that most folks wouldn't have any reason to know. That said, most folks reading this may know that I have been teaching survival professionally since 1993—that's thirty years. I've written almost a dozen books, done one film, and over fifty television shows all in the survival and special ops space.

When I first started writing books, it was only because an editor for Perseus, Greg Jones, saw my one-off special on Amazon Survival about what an expert would do in the same place as the survival story on Yossi Ghinsberg, which became a film titled *Jungle* with Daniel Radcliffe playing the lead role. Discovery had made a great show called *I Shouldn't Be Alive* and wanted to make "bookend" stories with experts put in the same situation.

I was asked to do the show *Science of Survival*, and that's where the publisher saw me and then asked me to write a small book on fire. I did. He loved it so much, he asked me, "What else can you write about survival?" To which I answered . . . everything! So, he let me write whatever I wanted, and when I was done, he decided to keep it all in. That became the *Green Beret Survival Manual*. It was not only a bestseller; it was rated as best in class by both the *Guardian* and *Kirkus Reviews*.

At one point, when the publisher asked me to make a thinner book for field use, they thought they'd stop printing the big book. The thinner handbook was water-resistant and had a soft cover, but it was literally only half the book. I found it harder to edit my fat book than to write the whole big one.

Anyway, they did stop, and then that original big book was selling on Amazon aftermarket for $2,000 USD. Mainly because it had good info that most other survival books didn't cover, like war, famine, disease, nukes, coups, revolutions, amputations, and cannibalism. So, they brought it back. It is now at home with Skyhorse, the same folks bringing you this one.

From that humble beginning, I ended up doing more and more television work, originally as a subject matter expert, then eventually, they asked me what shows I wanted to do. I created *Naked and Afraid, Dude, You're Screwed, Escape, Lost Survivors, Elite Tactical Unit,* and *One Man Army*; then I was blessed to do a show with my wife, Ruth England, and that show was called *Man, Woman, Wild.*

It's of interest to me that back then, Discovery didn't even think social media was important, so we made our own pages online. It became something of an issue a few years later, but that is a story for another time. The key point here is that my objectives as a teacher on that show were to make sure we didn't reinvent the wheel and do the same old things, but also to teach many new, unique, and different points during each show so that it added to the pantheon of survival shows with new knowledge.

My goal on our show was, first and foremost, to teach folks the key decision-making criteria for whether to stay or go. And then, different ways to do the eight pillars of survival for those type of environments. For example, to teach the four primary pillars of survival as Food, Water, Fire, and Shelter, all with a shifting priority based on current circumstances, and then to delve further into the four secondary pillars of Medical, Signal, Navigation, and Protection or Security/Tools/Weapons.

Secondly, of these, having taught survival for so many years, I knew the single most important actual survival skill is fire. First aid to stay alive and navigation to get out of the survival situation are also critical, but navigation can be aided by GPS and first aid can be assisted by EMS—but no one is going to make a fire for you.

With all that said, I wanted to share a few records we broke during our time making survival shows before I close this diatribe.

During the two seasons of *Man, Woman, Wild*, we broke many reality TV barriers, such as a full episode at sea, a real urban survival episode no one else had done, and more than one and a half segments in a cave—that means two commercial breaks. We tapped out for real in the desert sun and Alaska snow, showing the truth that heat and cold kill if you're not ready. I was also forced to do a real enema on a show, traversed a real minefield, scaled down a true death-defying razor cliff over nine hundred feet high, and many more things.

Finally, the factoid I am most proud of, other than the fact my wife and I survived all that and our marriage, is the fact that we taught twenty-two different ways to start a fire on twenty-two different episodes. That's pretty neat and was harder than it sounds.

PERSONAL

On a personal level, I have worked as a street paramedic and dealt with fires while supporting the local fire department. I have seen some sad things I can't unsee. I would strongly encourage everyone to take this chapter very seriously. Of all the things I have witnessed, other than war, the victims of burns and fires truly haunt me the most.

Now, to end this little piece on a positive note. Another way we broke records on reality TV was in our fire episode. We have some amazing humans that are called "smoke jumpers." They are firefighters who jump into forest fires in heavy equipment to protect themselves for tree landings, then they carry everything on their back to fight the fire. It is for this reason, the CIA actually hired some to work with special forces in Vietnam.

I did one CIA contract in my early days, and they brought some smoke jumpers to the war. I have a great deal of respect for them, from that experience and my knowledge that many folks die each year caught in out-of-control wildfires. So, I wanted to find a way to give some hope and provide real solutions to implement in an emergency.

To get after this, I did what we have always done in special forces—we go to the truest, best experts in the subject, usually civilians, as they can focus on that one singular problem set, backed by educated experts and scientists seeking how best to save lives.

We spoke with subject matter experts in firefighting from the federal, military, and state fire departments to come up with a viable formula for how to potentially survive a fire.

Firstly, we got as close to water as we could. If big enough, we would have gone in and used a straw to breathe through with a cloth on top to filter out ash. But we only found a small hole at the time. So, we worked with what we had.

We spoke with all three fire departments, and they all recommended about three to four inches of wet soil or mud may be enough to shield from heat and fire. So, we tried to triple it at twelve inches of packed mud, as we planned to be in the shelter when the fire got to us. We knew we couldn't outrun the fire, and it was too wide an area to pivot out through a flank. So, we had to take other measures.

We knew we had a few hours, so the first thing we did was pick a good, low area next to a water source. We made our own firebreak the best we could, then we made our own fire shelter. We made holes for our faces to face into and hopefully have enough space to get air not filled with smoke; we soaked ourselves and held cloths to our faces and prayed. Thankfully, it worked. I don't recommend it, but if it's all you've got, it's worth a shot.

FAMILY

When it comes to family, most folks will say it is what they value most, and that's as it should be. And depending on where you live, what your real chances are of a wildfire or a house fire, you may want to get personal gear for each member if you can afford it, can store it, and can replace it or rotate it at least every decade.

I'll include lists of gear the pros use, so you can either get those, or let that be your gold standard as you seek out what works for you.

NEIGHBORS

One thing for sure when it comes to fires, your neighbors matter. If it's your home that caught fire, it could spread to their home. Or vice versa. And in either case, it's always better if you can rely on each other.

HOME

There are many things you can do to help make your home more ready to deal with a fire should it come calling. These may be cost-prohibitive for many, but by knowing what is out there, you can find your own best way to develop each of these potentially mitigating options.

Starting from the top down, let's look at your roof. Most tar and clay shingles or tin roofs are fairly fire resistant in that they don't ignite with sparks and embers. That doesn't mean they won't burn in a high-intensity fire, but they give you some break.

However, having gutters full of debris can catch fire, so keeping them clean or screened is a good fire prevention technique. Likewise, you will want to put a metal mesh screen on all your vents that face outdoors, and that goes for any basement or crawlspace vents. Anywhere an ember can get in, it can ignite a fire, so put up vent screens.

Likewise, you can double-pane your windows and use fire-retardant curtains. These are not guarantees of protection but they certainly help with potential survivability.

There are also paints and sidings that do better against fires than others, and of course bricks and stones are better in this area. But we're not here to rebuild your home, so enhance your chances by shoring up what you can, where you can.

And in all this, don't forget to get all the bells and whistles for inside the house. Have fire detectors, carbon monoxide detectors, safety equipment, fire extinguishers of all sorts; practice with them, inspect them, and keep everything up to date and ready for action. In a survival situation, be sure to fill your tubs and sinks in case you need the water to fight the fire or to try to take refuge in—but that is a desperate measure, and escape would be preferred.

So, here's a synopsis of what you can do during a fire at home. Here are some things that can help protect your family in a fire, or to prevent fire:

1. Smoke detectors: Install smoke detectors on every floor of your home and in every sleeping area. Test them regularly and replace the batteries at least once a year.
2. Fire extinguishers: Keep fire extinguishers in several locations in your home, such as the kitchen and garage. Make sure everyone in your family knows how to use them.
3. Escape ladder: If you have multiple levels in your home, consider keeping an escape ladder in each bedroom to allow for a quick exit from upper floors in case of a fire.
4. Emergency kit: Prepare an emergency kit with essential items such as water, nonperishable food, a first aid kit, and a flashlight. Keep it in an easily accessible location.

5. Safe meeting place: Establish a safe meeting place outside of your home where everyone can gather in case of a fire. Practice fire drills to ensure everyone knows what to do.

6. Fire-resistant safes: Consider investing in fire-resistant safes to store important documents, valuables, and sentimental items.

7. Fire-resistant clothing: Keep a set of long pants, long-sleeved shirts, and boots made of fire-resistant materials in your home. This can help protect you if you need to evacuate in a hurry.

8. Clear clutter: Keep your home free of clutter and combustible materials, such as newspapers and cardboard boxes. Store flammable materials, such as gasoline and propane, in a cool, dry place outside the home.

9. Maintain electrical systems: Have your electrical systems inspected and maintained regularly by a licensed electrician. Avoid overloading outlets and using frayed or damaged electrical cords.

10. Install fire-resistant roofing: Consider installing fire-resistant roofing materials, such as metal or tile, to reduce the risk of fire from outside sources.

11. Maintain landscaping: Keep trees and shrubs trimmed and away from your home to reduce the risk of fire spreading to your home.

12. Install fire-resistant siding: Consider installing fire-resistant siding materials, such as brick or stucco, to reduce the risk of fire from outside sources.

13. Plan and practice escape routes: Plan and practice escape routes with your family in case of a fire. Make sure windows and doors open easily and are not blocked.

14. Block the attic vent screens where embers can enter and start fires inside. Protect exposed wood under the eaves of your home.

15. Cover windows to prevent infrared heat from igniting materials inside house. Remove flammable materials from around your house.

16. Set up adjustable Rain Bird sprinklers to cover the sides and eaves of your house. Fireproof paints, gels, coatings, and materials.
17. Fire alarms, detectors and response systems.
18. If you have a swimming pool, get a swimming pool fire pump.

Remember, the most important thing to do in case of a fire is to get out of your home as quickly as possible. Be prepared, stay informed, and practice fire safety measures regularly with your family.

YARD

This is actually one of your best lines of defense. So, make your home defensible against fires. There are various recommendations for zones sizes, but all the experts agree on zones for planning. I will keep it simple and use standard military metrics. When we're on a weapons range, we consider the fifty-meter target to be the most dangerous as it is the closest. The next level out is a hundred meters, and while we do have a one-hundred-fifty-meter target, I will use the two-hundred-meter zone here.

But let's convert these to feet for city folks since most don't work in meters or yards, especially when it comes to modern urban yards. So, let's have two simple zoning plans:

Urban Fire Zoning Plan:
Zone A: 0–50 Feet
Zone B: 50–100 Feet
Zone C: 100–200 Feet
Rural Fire Zoning Plan:
Zone A: 0–50 Yards/Meters
Zone B: 50–100 Yards/Meters
Zone C: 100–200 Yards/Meters

Now, for your Zone A, clear anything flammable such as wood piles, any junk or clutter, fallen trees, foliage, propane tanks, etc. It is best to use rocks, stones, sand, cement, etc., for the area immediately around your home. You may want to water your lawn if you have a sprinkler system. But take into consideration that this will reduce water pressure elsewhere that firefighters may need.

For Zone B, it is better if you don't have highly flammable plants like pines and others that have high concentrations of resins or saps that can burn. Remove all dead, fallen, and decaying trees or shrubs that can catch fire.

For Zone C, it comes down to landscaping. If you do this with fires in mind, you should seek to select plants like succulents that don't readily burn, and likewise, you'll seek to avoid those plants that are more combustible. Either way, all of this comes down to you making your last defense, so make it the best you can.

VEHICLE

There's not a whole lot you can do to fireproof your car, but that shouldn't stop you from trying. I'm sure some folks out there have already invented some type of fireproof mobility option, and it could be good fun to do some research and see what is out there, and if there is anything you can do to hedge your bets.

For sure, you should have some standard basics for dealing with fires, even if it's just a car fire. So, a proper fire extinguisher for car, gas, oil, and electric fires is a good start. Some water for many reasons, but for helping with fire it is also good. A fire-resistant blanket can help put someone out who may have been in an automobile accident and caught fire, or just a simple first aid kit with some treatment capability for burns.

If you are caught in the middle of a wildfire while driving, it's important to stay calm and take action quickly to protect yourself and your family. The items to have in your vehicle for heat are applicable here (masks, emergency kits, protective clothing, blankets/tarps, radios, maps, flashlights, bottled water, etc.)

Remember, the best way to protect yourself from a wildfire is to stay informed, follow evacuation orders, and have a plan in place. If you are caught in a wildfire while driving, stay calm and listen to local authorities. Follow their instructions and evacuate immediately if necessary.

STATIC PLANNING

First and foremost, if you think you may be in even a remote chance of being caught in a wildfire, the best thing is always to avoid it by escaping early. But not everyone has that option, so they have to make a plan to shelter in place.

Much of what we discuss here can also be considered for your standard home fire plan as well. In short, if the fire starts in your house, get out of it, and if it's outside your home, stay in. But have a plan for your family to deal with home fires, rally points, who has what responsibilities, what are the priorities for what to take, and what to leave.

You could save a life just by keeping all things precious, be it papers and photos, gems or gold, or anything else, in a fireproof safe. The best kinds of safes are also waterproof, so you have double protection and a lot less to worry about.

Keep all electronics backed up in the cloud or on a hard drive, consider safety deposit boxes for those as well as for any other valuables, collectibles, or items of personal importance to you and your family.

Have a plan for first aid, especially for when the primary caretaker becomes the victim, and try to be ready to help others if it spreads. It is a good idea for rural dwellers to plan on care to sustain any burn victims long enough for EMS to arrive or for you to get them to help, however far that may be. If you absolutely must remain in place, then here are some considerations.

If you must stay put and let a wildfire pass over you, it's important to take precautions to stay safe. Here are some things you can do:

1. Find a safe location: If you are unable to evacuate, find a location that is as far away from the fire as possible, and preferably uphill or upwind. Seek shelter in a building if possible, but avoid basements or other enclosed spaces where smoke can accumulate.

2. Create a fire-resistant barrier: Use fire-resistant materials to create a barrier around your home or location. This can include removing any flammable materials within thirty feet of your location, keeping your lawn watered and well-maintained, and creating a defensible space around your location.

3. Stay hydrated: Drink plenty of water to stay hydrated, as the heat from the fire can quickly dehydrate you.

4. Protect yourself from smoke and heat: Use an N95 mask to protect your lungs from smoke and ash. Wear long pants, long sleeves, and boots made of fire-resistant materials. Use wet towels or blankets to cover your body and protect yourself from heat.

5. Stay low to the ground: If the fire is approaching, stay low to the ground to reduce exposure to smoke and heat.

6. Stay informed: Listen to local news and emergency services for updates on the fire and any evacuation orders.

Remember, staying put during a wildfire should only be done as a last resort. If you are able to evacuate, it's always safer to do so. Be prepared, stay informed, and take precautions to stay safe.

MOBILITY PLANNING

Here's where the vehicle planning can make or break your survival efforts. Have a planned route that takes into consideration *all* aspects of wildfires. This sounds daunting—and it can be—and it will never be foolproof, but a plan is never a guarantee, it's merely a start point for reality.

Consider the prevailing winds where you live. Do you have trade winds? Do they change direction with the time of year, or time of day? Many folks don't realize how wind can draft upwards in mornings as the sun heats the earth, or that they can flow downward in evenings as the sun leaves the surface to cool. You may have seasonal winds, or not, but the key when fire planning is to get familiar with your winds, daily and seasonally, as they will 100 percent impact you in the case of a large-scale fire.

Also, look for lower-lying routes, as fires and smoke will rise. Seek open areas to avoid being channelled or trapped. Keep in mind water sources and crossings. Have a place to go, a few backups, and a plan in case anyone gets hurt or lost at any point of the journey. If you have to plan a breakout journey through a wildfire, you might want to think about some of these things.

If you must plan a dangerous car journey through a wildfire to escape, here are some things you should have and do to increase your chances of survival:

1. Listen to local authorities: Before attempting to escape, listen to local authorities for updates on the wildfire, road closures, and evacuation orders. Follow their instructions and be prepared to change course or turn back if necessary.

2. Prepare your car: Make sure your car is in good condition and has a full tank of gas. Close all windows and air vents to keep smoke out of the car.

3. Pack essentials: Pack a go bag with essential items, including water, nonperishable food, a first aid kit, a portable radio, and extra batteries. Bring blankets or tarps to cover the car in case of falling debris.

4. Choose the right route: Plan your escape route carefully and choose the safest and shortest route possible. Avoid narrow roads and dead-end streets.

5. Stay low: While driving, stay as low as possible to the ground to reduce exposure to smoke and heat. Keep the air conditioning on and use the recirculation setting to avoid drawing in smoke.

6. Use headlights and hazard lights: Turn on your headlights to increase visibility and use your hazard lights to signal to other drivers that you are in distress.

7. Be aware of other drivers: Be aware of other drivers on the road and watch out for emergency vehicles. Stay in your lane and avoid sudden movements.

Remember, attempting to drive through a wildfire should only be done as a last resort. If possible, always evacuate and seek shelter in a safe location. Stay informed, be prepared, and prioritize your safety above all else.

INTRAPERSONAL

On the intrapersonal realm, I recommend you take a moment and look up burn victims. It will scar you, mentally and emotionally to be sure, no getting around it. But if you are to face it, you must know what you face. And the more you look at post-burn surgeries, repairs, plastic surgery, prosthetics, and others who have fully or mostly recovered, the more it helps you have hope, that if you have to face a potential burn to rescue a loved one, there is a chance you may survive and can still have a good quality of life. But this is how I operate. It may be too harsh for some. Either way, to make peace with the potential of being burned is important for you to overcome fear and do what needs to be done when it matters most.

MISCELLANEOUS

Included here are some neat and novel things I found that may or may not be of interest or service. There are several new technologies that can help save your life if caught in a wildfire:

1. Fire-tracking apps: There are now several apps available that provide real-time updates on the location and severity of wildfires. These apps can help you stay informed and make decisions about evacuation and safety.

2. Personal fire shelters: New lightweight personal fire shelters are available that can be quickly deployed to protect you from flames and heat.

3. Fire-resistant materials: Advances in materials science have led to the development of new fire-resistant materials that can be used in construction, clothing, and other products. These materials can help protect you and your property from wildfires.

4. Drones: Firefighting agencies are now using drones to gather information about wildfires and monitor fire behavior. Drones can be equipped with thermal imaging cameras to detect hotspots and track the movement of fires.

5. Smoke-filtering masks: New high-tech, smoke-filtering masks are available that can filter out harmful particles and help you breathe more easily during a wildfire.

6. Early warning systems: Some companies are developing early warning systems that use sensors to detect the presence of wildfires and alert people in the area. These systems can provide valuable time for evacuation and preparation.

Remember, while new technologies can be helpful, they should not be relied upon as the only means of protection during a wildfire. Always follow local authorities' instructions, evacuate when necessary, and prioritize your safety above all else.

SECTION II
COLD AND ICE

When I shiver I do so not in fear but in the stark reality that icy fingers are testing my resilience to the cold.

3

Ice Ice Baby

COLD—THE SITUATION

As a meteorologist, for many years and with extreme weather reporting part and parcel of the job, I have occasionally thought about which of the two temperature evils I'd choose over the other to succumb to if I had no choice—extreme heat or extreme cold. Of course, it's a rabid Hobson's choice and one I sincerely hope I shall never have to suffer, nor you for that matter! But I do have a preference.

In my time, I have personally experienced both blistering heat and extreme cold. My big heat was endured while on a visit to the Chott el Djerid, an endorheic salt lake in central Tunisia, during high summer, which I can only describe as being instantly debilitating. My big chill was encountered during an evening foray on a winter holiday in a mountain hideaway within central Norway, equally debilitating, and strangely not so different in my head to the impact of the salt lake oven. Those extremes were somewhere around 118°F (48°C) in the heat, and minus 22°F (minus 30°C) in the freezer.

On the evening of that extreme cold, my young middle son, Christopher, and I dared each other to run out of the comfort of the hot sauna and dive into the deep, frozen snow. It lasted for no more than two minutes as we joyously frolicked about before returning to the refuge of the sauna with seriously numbed feet, frozen fingers, tingling ears, and blue noses.

That might seem somewhat fanciful, even temporarily idyllic, and indeed, in truth it was. But in that couple of minutes of deep-freeze exposure, and even with a head start having the warmth of the sauna both on and within our bodies, the cold immediately took over and made it far too risky for us to continue our escapade. You see, our bodies cooled down so fast that our inner heat immediately concentrated itself within our cores for protection, leaving our exposed feet, fingers, noses, and ears drained of blood. Our extremities would have been at risk of frostbite had we hung around in our near-naked state for much longer.

It doesn't matter whether we are talking about exposure to the cold in a polar region, on a mountain peak, in a sea, lake or river, at a football game, waiting at a bus stop, or homeless and sleeping out on a park bench; if it's cold enough for long enough, that cold will find its way into your veins and capillaries sooner or later, or more accurately, your inner heat will find its way out.

In open-air situations, exposed skin can be prone to frostbite when the air temperature slips below freezing point. It's not an immediate thing, but once you notice color changes on skin surfaces, it's a pretty good indicator that something much worse could be around the corner. The longer a person continues to be exposed within the freezer zone, the colder the body will become and the more ingrained the frostbite is likely to be, as blood exits extremities heading inwards toward vital organs.

Frostbite isn't to be welcomed, of course, but a more serious condition (with or without the frostbite) is reached when the human body ventures into a hypothermic state. That point is around 95°F (35°C), remembering that humans ideally function between 97°F (36.1°C) and 99°F (37.2°C). It will vary dependent on the status of the individual and the surrounds, and those with more layers of fat will tend to last longer; if the hypothermic state continues for up to an hour (and it could be less), then a complete shutdown of bodily functions can occur for anyone, no matter the stature. Death then becomes virtually inevitable.

Mykel will delve a little deeper into the negative signs to be aware of, and how best to protect and survive if you find yourself compromised by the cold. But for now, I would like to add two meteorological elements to the story that could compromise your situation even further at such times: wind and moisture. Let's just say that neither of these two will be your buddy when it comes to surviving the cold. As soon as there is a wind blowing up against you, even a slight wind, there will be an added wind chill. We've all been there when the wind has been blowing on an otherwise warm day and it simply feels less warm or even chilly. On a cold day, that wind can be bitter, particularly on exposed skin surfaces, such as faces and hands. Wind chill in critical survival situations is a demon, and the harder it blows (especially if it is of polar or semi-polar origin wind), the greater the wind chill will be. It's not that the air temperature falls further because of it (it doesn't), but the wind will draw heat off our bodies at a faster rate than would otherwise be the case. Exposed skin will suffer the consequences much more than skin under wraps, but wind chill will penetrate as far as it is allowed to.

Back in 2001–02, Environment Canada and the US National Weather Service created the first wind chill chart, which is widely used today in expressing the relationship between various wind speeds and a range of air temperatures on exposed skin. Many versions of such charts are available online but I'll leave you with a couple of examples to put into your back pocket. The first is a light breeze of 10 mph and an air temperature of 32 (°F)—the freezing point of water, which converts to an equivalent wind chill value of 24 (°F), or minus 4.4 (°C) if you prefer it that way. You see, it doesn't take much. Now let's ramp it up (or in this case down) with a wind speed of say 30 mph and an air temperature of 20 (°F) or minus 6.7 (°C). Then the wind chill conversion results in an air temperature of just 1 (°F) or minus 17 (°C). That's serious stuff!

And then there is moisture. Being wet, or getting wet, is another demonic element. I'll keep this simple; a wet body from whatever means (including being immersed in freezing sea water or from just mere sweating) will evaporate heat from your body between five and twenty-five times faster than the surrounding air will. In essence, being drenched or immersed in water, even if the air temperature is above freezing, will chill the body alarmingly quickly. Have you ever sweated and then stood around on a cool or cold day? Yep, that cold feeling running down your spine and consequential shivering is a sure sign that you are giving up heat to the air around you faster than you might like. You may have also observed, on television or in the flesh, those steamy outflows from sweat-laden bodies of male and female athletes on a cold day. You are simply observing physics in action that entails body heat being lost to the surrounding air in drafts of body moisture.

Finally, you might think that with all the talk about global warming and record-breaking heat, the risk of a death in the freezer might be somewhat reduced. After all, it's a temperature one-way street, isn't it? Well, no it's not, actually! That word *chaos* comes back into the discussion here because a warming world will eventually dislocate ocean currents, such as the Gulf Stream for example, which makes northwestern Europe in winter that much milder than it would otherwise be for its latitude. It will also act upon the polar vortexes, likely making them weaker and more erratic, feeding freezing cold air, instantaneously crippling ice and deep snow out of the Poles and into highly populated conurbations. Bottom line, a warming world can also mean spells and periods of sudden catastrophic freezing during the winter season and either side of it. Certainly, it is prudent to take nothing for granted!

As he has expertly done with extreme heat, in any and all deep freeze situations, Mykel will advise both you and me on how to survive when your body thermometer might otherwise give up the icy ghost!

COLD—THE SURVIVAL

Jim's done a fine job on the weather science, now let's look at applied science for dynamic solutions. Being caught ill-prepared for extreme cold weather has claimed many tens of thousands, if not millions, of souls over those same millennia. When it comes to heat, water and shade are your main actions, but when it comes to cold, there are many tools in your arsenal. But it comes down to simply getting out of the cold or staying warm while in it.

There are two kinds of cold weather survival situations—prepared and unprepared.

Prepared just means you live there or traveled to a cold weather region intentionally. If you live there, you've either grown up with cold weather as part of the seasons and mostly have the gear you need for daily life year-round.

For others, they travel to visit family, friends, ski or climb, or just generally planned to go somewhere cold or in winter season. The assumption is that folks have common sense and pack accordingly.

However, folks can find themselves stranded in cold weather unexpectedly, or be caught in some unusual weather. For example, folks in Siberia endured a minus 50 temp that had even their hardiest shivering. Let's dig in to all the aspects of prepping, dealing, and surviving cold weather.

First, let's take a look at some of the coldest places on earth to set a baseline of what our normal extremes look like, so you can use that as a metric should we begin to see extremely cold weather that is worse than anything recorded before. The coldest places on Earth are primarily located in the polar regions, specifically in Antarctica and the Arctic. Some of the coldest places on Earth include:

1. Vostok Station, Antarctica: This research station holds the record for the coldest temperature ever recorded on Earth, at minus 128.6 degrees Fahrenheit (minus 89.2 degrees Celsius) in 1983.

2. Dome A, Antarctica: This is the highest point on the East Antarctic Ice Sheet, and temperatures can drop below minus 100 degrees Fahrenheit (minus 73 degrees Celsius).

3. Oymyakon, Russia: This village in Siberia is considered the coldest inhabited place on Earth, with temperatures as low as minus 90 degrees Fahrenheit (minus 67 degrees Celsius) recorded in the winter.

4. Verkhoyansk, Russia: Another Siberian village, Verkhoyansk is known for its extreme temperature swings, with a record low of minus 93.6 degrees Fahrenheit (minus 69.8 degrees Celsius) and a record high of 98 degrees Fahrenheit (36.6 degrees Celsius).

5. Eureka, Canada: This research station in Nunavut, Canada, experiences bitterly cold winters, with temperatures sometimes dropping to minus 50 degrees Fahrenheit (minus 45 degrees Celsius).

These are just a few examples of the coldest places on Earth, but there are many other locations in the polar regions where temperatures can be incredibly low.

Now let's look at some of the most extreme cold storms ever recorded, so you can determine if you are ever in one worse than these, you will understand the gravity of the situation so you can correctly calibrate your courses of action countermeasures.

The coldest storms ever recorded are typically associated with extreme cold weather events in the polar regions. Here are some of the coldest storms ever recorded:

1. The Superstorm of 1993: This was a massive storm system that affected much of the eastern United States in March 1993. The storm brought record low temperatures and heavy snowfall, with some areas receiving more than four feet (1.2 meters) of snow.

2. The Siberian Express of 1983: This storm system affected much of the central and eastern United States in January and February 1983. The storm brought record low temperatures, with some areas experiencing temperatures as low as minus 80 degrees Fahrenheit (minus 62 degrees Celsius) with wind chill.

3. The Winter Storm of 1996: This was a severe winter storm that affected much of the eastern United States in January 1996. The storm brought record snowfall and frigid temperatures, with some areas experiencing temperatures as low as minus 40 degrees Fahrenheit (minus 40 degrees Celsius).

4. The Polar Vortex of 2014: This storm system affected much of the United States and Canada in January and February 2014. The storm brought record low temperatures, with some areas experiencing temperatures as low as minus 50 degrees Fahrenheit (minus 46 degrees Celsius) with wind chill.

Now let's look at what to do if the extreme cold causes you harm directly.

COLD WEATHER MEDICAL ISSUES AND TREATMENT

First and foremost, when talking about the cold, we have to discuss the illnesses and injuries that can be caused by extreme cold weather on the human body—they include: how to recognize them, how to treat them, and ideally how to prevent them in the first place.

Here are some basic considerations to help you fight off the cold.

Stay Hydrated. It's easy to not want to drink when it's cold, but it's imperative that you stay just as hydrated in extreme cold as you have to do in extreme heat. Just think of water as the body's oil, like a vehicle; if it gets too thick and viscous, then it's harder for it to do its job. That means you get colder, and a lack of proper hydration in the cold is setting you up to be a cold-weather casualty. Hot coffee and tea are okay, but they have caffeine, which is a diuretic—meaning it makes your kidneys process the water out of your body faster, increasing dehydration.

You can drink these, but you need to counterbalance with a full glass of water after each cup of caffeinated beverage. I tend to drink plain hot water frequently during the cold as the best way to stay hydrated and actually keep your core temperature up.

Extreme Hydration. It is commonly known that one should not eat snow when surviving, but this, like many other myths around survival, has merit and validity as a general concept, but also fails to impart potentially lifesaving information for when there are exceptions to the rule.

I believe there are times when you can eat snow, but you must understand when and how. First, you should understand how the body operates in regards to temperature. Most folks know that the normal body temp is 98.6 degrees Fahrenheit. If your body temp drops to 95 degrees, that is usually when it shuts down; you can no longer function, and the end result is that you freeze to death. That's only a drop of three degrees. And that's why it is absolutely vital that you stay warm and dry to prevent hypothermia.

In general, it is best to not eat snow when in the cold. If you're freezing already, do not eat snow as it only serves to drop your core body temperature even more, and the effort your body has to make to warm the snow as it enters your esophagus actually makes it expend more energy, leading to more dehydration.

Some background for a point of reference: I spent my first four years in special forces serving on a winter warfare team. I thought snowshoes and igloos were great things and enjoyed the cross-country and downhill skiing, among other things. During that period, I learned from the old-timers a nifty trick to stay hydrated on the move.

Believe it or not, when you're carrying a 120-pound backpack, uphill in three feet of snow for miles at a time, you can actually overheat. During these times, if you're not able to grab your water bottle or stop to heat some water, grabbing a handful of freshly fallen, undisturbed snow with clean hands will actually help you hydrate; the cooling effects that in other circumstances could kill you, actually help you avoid overheating and losing more sweat and becoming more dehydrated. Try this sometime when you're in the snow doing some demanding physical activities and see how it cools you down while hydrating you, too.

Keep Dry. Once your clothes become wet from sweat or precipitation, they lose some of their ability to retain heat and become a great conductor for pulling heat away from your body. In short, getting wet in the cold can be a killer unless you can take action to get warm and dry pretty quickly. A good change of clothes is often the best immediate measure, followed by a fire or getting out of the cold altogether.

Cover Up. The more body surface area you can cover up, the more heat you'll be able to retain. Always have gloves, hat, and scarf with your winter coat and vests, earmuffs, and of course, all the usual cold-weather tops, pants, and footgear. A key for clothing in cold weather is not to wear things too tightly. Loose and spacious clothing traps air and body heat helping to keep you dry and warm.

Keep Moving. If you're getting cold, and the usual options like fire and shelter are not available, move around to keep the body generating some heat.

Eat Well. Moving expends energy, so keep replenishing your food stores by eating. This is one of those times where carbs and fats are your friends. Inuit Eskimos consider whale blubber and seal fat ideal foods.

Seek Shelter. If you start to shiver, the process of hypothermia has already begun. If you can't get out of the cold, then you should consider stopping and making a shelter and a fire. This work will keep you warmed up, and once in a shelter, that will warm you considerably, even if you can't get a fire going. And of course, fire is your friend in the cold.

COLD WEATHER ISSUES BY CATEGORY

Water/Ice: Water freezes at 32 degrees. It's important to have an understanding of what happens when water freezes. Primarily, the water expands, and that means anything containing water that doesn't have enough space to expand will likely rupture. Canteens and pipes come to mind.

Reading Ice. It's also important to be able to read the signs of ice in order to understand what you can and can't do. For example, when a lake is solid white, not covered in snow, there's a good chance it is solid enough to traverse or cross. And when driving, it's vital to be able to discern if there is black ice, or that thin layer of wetness on tarmac that has frozen, giving a deadly slick surface on roadways.

Air/Wind: When it gets really cold, high winds exacerbate the effects on human tissue. For example, it can be a solid 32 degrees, but if the winds are howling and gusting up to twenty knots, then that has the effect of making it exponentially colder on tissue. So much so, that even a quick run of fifty yards from the cabin to the outhouse, with exposed skin for only a few seconds, can result in frostbite. Untreated, that could lead to gangrene, loss of tissue, digits, limbs, and even life.

Pets/Animals: An often-overlooked concern when preparing for cold weather is how to protect your pets and animals. They'll need shelter with heat, they'll likely need coats or blankets, and perhaps even booties or socks. If exposure to high winds is expected to be a regular thing, you may even consider goggles. They'll need all the things people need as well, such as plenty of food, water, and warmed up for both is good, too. Remember, we use salt to remove ice, and uncovered pet paws can get salt in them, cracking their skin and severely risking frostbite. All this also applies to any livestock. Yes, they can fare better than us in the cold, but they still do get cold like us, and they can perish in extreme cold weather.

COLD WEATHER CHALLENGES FOR HOME

Firstly, let's talk about how you might tell if a cold snap is coming. This will give you a better chance to make emergency preparations. Cold snaps can be difficult to predict, but there are several signs to watch for that may indicate a sudden drop in temperature is on the way. Here are some of the most common indicators of a cold snap:

1. Sudden changes in air pressure: A rapid drop in air pressure can indicate the approach of a cold front and a potential cold snap.
2. Changes in wind direction: A shift in wind direction from the west to the north or northeast can signal the arrival of colder air.
3. Cloud formations: The appearance of thick, dark clouds can indicate a cold front approaching.
4. Precipitation type: If rain suddenly turns to sleet or snow, it could be a sign that a cold snap is on the way.

5. Animal behavior: Some animals, such as birds, may flock together and migrate south in response to a cold snap.
6. Weather forecasts: Checking the local weather forecast can provide valuable information about temperature trends and the likelihood of a cold snap.

It's important to note that these signs do not always indicate a cold snap, and sudden drops in temperature can still be unpredictable. It's always a good idea to be prepared for cold weather by dressing in warm layers and having emergency supplies on hand.

Secondly, it's important to make sure you have multiple ways to keep warm at home in case of extreme cold weather. If power goes down, it's good to have a generator. It's also good to have a gas stove for outdoor cooking and heating of water, as well as a charcoal grill for when that fails. Some folks have kerosene heat, and it's good to have a plentiful supply other than what's in the tank.

We then have our camping stoves as backups to that, we have a small wooden stove for when all else fails, and a good log pile of wood to see us through a very long time. (Note: never use any gas products—stoves or generators—indoors, and if you must, then it is critical to keep the area very well ventilated). Those gas products emit a byproduct called carbon monoxide, and it kills about a thousand Americans every year. So, it's good to have a carbon monoxide detector installed in your home.

For your wood stoves, make sure you have a good screen to prevent embers from starting a fire. Never use gas to start a wood fire, as it can explode a fireball in your face, and don't close your damper while the ashes are hot. Don't burn charcoal indoors, either. Space heaters are great so long as you have electricity or a generator to run it.

Before we get too deep in COAS for extreme cold weather, let's look at a quick overview. If you live in an area with cold winters, it's important to prepare your home for the colder temperatures to ensure that you and your family stay warm and comfortable throughout the season. Here are some of the best things you can do to cold-proof your home (as with home preparedness and maintenance for heat, these are good home protocols in general):

1. Insulate your home: Insulation helps to keep the warm air inside and prevent cold air from seeping in. Make sure your attic, walls, and floors are well insulated.
2. Seal air leaks: Air leaks around windows, doors, and pipes can allow cold air to enter your home. Use weather stripping and caulk to seal these leaks and keep cold air out.
3. Upgrade your windows: If your windows are old or single-pane, consider upgrading to more energy-efficient models. Double- or triple-pane windows can help keep cold air out and reduce heating costs.
4. Maintain your heating system: Regular maintenance of your heating system, including cleaning and replacing filters, can help keep it running efficiently and prevent breakdowns during cold weather.
5. Use a programmable thermostat: A programmable thermostat can help you save money on heating costs by automatically adjusting the temperature when you're away from home or asleep.
6. Install a storm door: Adding a storm door to your entryway can provide an extra layer of insulation and help prevent drafts.
7. Use draft stoppers: Draft stoppers placed at the bottom of doors and windows can help prevent cold air from entering your home.
8. Consider a generator: In areas with severe winter weather, power outages can be common. A generator can help keep your home warm and powered during an outage.

For your pipes, to keep them from freezing on an outside wall, let hot and cold water trickle or drip at night from a faucet. Open cabinet doors to allow more heat to get to uninsulated pipes under a sink. Don't let the temp get lower than 55 degrees. Drain and shut off the water system if you're going to be away (except indoor sprinkler systems).

If your pipes freeze, make sure you know how to shut off the water, in case pipes burst, and make sure the rest of the family knows how to shut it off as well. Don't try to thaw a pipe with an open flame or torch. Use a blow-dryer instead. And always be careful of the potential for electric shock in and around standing water.

Check the insulation in your attic, basement, and walls and for any pipes that may be subjected to freezing temperatures. Ventilation in both of these is important, too, as that can help reduce potential ice buildup. There are many different types of insulations from foams to tapes, so be sure to research what's best for your needs. There are many new contraptions out there like faucet covers, hole fillers, etc. worth considering for protecting your home against brutal cold.

Another step to consider for making your home better equipped and prepared for extreme cold weather is to conduct an inspection before winter sets in to look for weaknesses in your fortress.

You can hire a professional to conduct a survey for you, and this will ultimately help save heat and money. Or you can do what we call in the military, operator preventive maintenance checks and services, or PMCS. Start with the basics—check for drafts under all your doors and windows, and vents in the basement and/or attic. Seal any that you find. There are many types of products on the market, and these are always changing. You'll have to do your own research to find the solutions that fit your budgetary parameters and constraints.

Sometimes, you may need to consider updating appliances if yours are still serving but getting antiquated. Older appliances do not use energy as efficiently as more modern appliances like water heaters, and I have found that things tend to break when you need them most. Losing an old stove or boiler in the middle of a terrible winter storm makes for a fairly uncomfortable experience.

Also, if you don't have a modern thermostat for your home, it is worth the investment, as keeping the house on an even temp can make all the difference. There is a neat company with a very impressive array of sensor suites that can help tremendously in the mission of keeping your home finely tuned, the same as you do for your vehicle. The company is called LiveWell. I have no business connection to them; I just really like what they do and how they do it.

Some will also suggest lowering your water heater temperature a bit to save energy and reduce risks of mild skin burns going from extreme cold to extreme heat. And it's not a bad idea to make sure your gutters are clean so they can drain well and not get clogged, freeze, and then create an ice hazard for you. It is also a good idea to keep your attached garage closed to help retain some heat against the walls it touches.

Here is a packing list for items to keep in your car (in addition to those previously covered that you should generally have):

1. Warm blankets: Thick, warm blankets can help keep you warm if you become stranded in your car.
2. Extra warm clothing: Pack extra warm clothing, including coats, hats, gloves, and boots, to stay warm if you need to leave your car.
3. Nonperishable food and water: Pack nonperishable food items such as energy bars, jerky, and nuts, as well as water and other beverages.
4. Hand warmers: Hand warmers can provide a quick source of heat to help keep you warm.
5. Flashlight and extra batteries: A flashlight can help you navigate in the dark, and extra batteries ensure it stays powered.
6. Portable phone charger: In case your car battery dies, a portable phone charger can help you stay connected.
7. Ice scraper and snow brush: These tools are essential for removing snow and ice from your car.
8. Jumper cables: Jumper cables can help restart your car if the battery dies.
9. First aid kit: Pack a first aid kit with bandages, gauze, and other essentials.

10. Road flares: Road flares can help signal for help and warn other drivers of your presence.
11. Shovel: A small shovel can be used to dig out snow from around your car.
12. Sand or kitty litter: Sand or kitty litter can help provide traction if your car gets stuck in snow or ice.

By keeping these items in your car, you'll be well-prepared for any emergencies that may arise during an extreme cold snap. Make sure to periodically check your emergency kit and replace any expired or used items.

COLD WEATHER CHALLENGES TO FOOT MOBILITY

When I first joined the army in 1982, we had big, fat white rubber boots, nicknamed Mickey Mouse boots, as they made your feet look abnormally large like a balloon. They had a small valve to allow some moisture to escape, but generally they were only good for standing, not hiking miles through the woods at night as we often did.

They made for good waterproofing of your feet, which was important, but they could also make your feet sweat during forced marches. To counter this, we used another time-tested and -proven material called wool socks. These were great for wicking moisture away from your feet, but they tended to cause friction blisters, so we would often use silk knee-high stockings which also allowed air in your sock, giving another layer of warmth, and that's how we took care of our feet with frequent rest stops to take boots off, air out feet—yes, even in the cold—and to change socks if needed.

These days, there are far better materials and products. Lightweight smart wools, silk socks, Gortex boots, and Vibram soles all make for great protection of your feet. As with everything, everyone's feet are a little different, so it's worth taking the time to research before you buy, read reviews, then once you purchase, make sure you get out and test your kit. Nothing worse than dropping a few hundred on some great footgear only to find it doesn't work as advertised for you. It is not uncommon for improperly fitted footgear to add to your problems by causing blisters, immersion foot, or frostbite.

Another worthy mention here are those electric (and sometimes fuel) warming devices, such as hand warmers for gloves and feet heaters for wearing in your boots, or electric gloves, hats, and vests. These can be plugged into the wall A/C socket or use batteries. Either way, I'm a fan.

COLD WEATHER CHALLENGES TO VEHICLE MOVEMENT

Most people who live in cold weather areas know all too well how important it is to keep your transport road-ready for winter. But that is not always true. I once lived in Minnesota for a winter. I had just returned from an awful deployment in Africa and had an offer to work for the army there. It was my turn to have the lads for a school year as my ex and I had divorced on amicable terms and took turns having the boys live with us. We made a family decision to move there. I had an Arctic wolf as a pet, and it all seemed great. But once we experienced minus 30 degrees, even colder than Alaska due to the double-cooling effect of the mid continental cold thermal and the Great Lake's chilling effects, we determined that kind of cold was for a hardier people. We had nothing but love for the accented folks of Minnesota.

However, prior to that, I had lived for years in the deserts of Nevada. So, you could imagine my surprise when every single snowstorm, without fail, there would be cars with MN tags sliding off the side of the road; meanwhile, our NV-tagged vehicle never lost traction or control.

That was mainly due to planning, training, and preparation. We had snow chains to put on our tires, we kept salt for de-icing, we had all the scrapers and shovels, blankets, and even electric plugs to keep our engine blocks from freezing at night.

An important part of any military operation is good logistics and maintenance, so we made sure to keep all the fluids fresh and topped up. The car was serviced regularly, we always took care of the battery, and kept jumper cables if we needed a charge, traction slats if we got stuck, a tow cable to help or be helped, and of course a full winter camping complement good enough for the two lads, the wolf, and myself to last a full week.

So, in addition to having all the normal survival stuff you need for you and your loved ones, pets and potential extended family, you'll need the cold weather gear for your vehicle, too. I always kept a bag of salt as well as sand and cat litter for various situations where you needed to melt some ice or get some traction.

Two key things I recommend for your vehicle winter-readiness plan outside of keeping it in good repair with fresh fluids and a full tank, are heaters for the two key components—the engine and the battery.

They make oil and block heater plugs that keep your engine and fluids warm to make it easier to start (and prevent freezing). And there are battery maintainers that provide a small 1.5–2 amp-trickle charge to make sure your battery is ready when you need it.

COLD WEATHER GEAR FOR PERSONNEL

Heated hats, gloves, and jackets are my favorite starters for my winter wardrobe. Wear clothing loosely and use layers. It is a constant juggle of clothing on and off to keep regulating your core temp and balancing it with whatever the temp of the weather effects might be. It's good to wear something over your mouth and nose to conserve moisture and reduce cold entering your core body and reducing your temperature. And lastly, it is a good idea to have mittens to wear over your gloves to retain finger heat, allowing you to remove when you need to use your fingers but having protection on them as well. Make sure they are snug at the wrist to keep snow from getting into your gloves and onto your hands.

Remember, *your clothing is your first-line shelter*. Staying warm and protected in an extreme cold snap is crucial for your health and safety. Here are some of the best things to wear to stay warm in extremely cold weather:

1. Base layers: Wear a set of thermal underwear made of wool or synthetic materials, such as polyester or polypropylene. This will help trap body heat and keep you warm.

2. Insulating layers: Add one or more insulating layers on top of your base layers. This can include a fleece jacket or wool sweater.

3. Outer layer: Your outer layer should be a waterproof and windproof jacket or parka. Look for one that is insulated with down or synthetic fibers for extra warmth.

4. Warm hat: Wear a hat that covers your ears to help retain body heat. A wool or synthetic beanie is a good choice.

5. Scarf or neck gaiter: A scarf or neck gaiter can provide extra protection to your neck and face from the cold.

6. Warm gloves or mittens: Choose gloves or mittens that are insulated and waterproof to keep your hands warm and dry.

7. Warm socks and boots: Wear thick socks made of wool or synthetic materials, and insulated boots with good traction.

8. Face mask or balaclava: In extremely cold weather, consider wearing a face mask or balaclava to protect your face from frostbite.

Remember to dress in layers so you can easily adjust your clothing as needed. Avoid wearing cotton as it retains moisture and can make you colder. Instead, choose clothing made of synthetic materials or wool, which wick away moisture and keep you warm.

COLD WEATHER GEAR FOR SURVIVAL

Extended Cold Weather Clothing System Generation III (ECWCS Gen3) is what the conventional military is using these days. Special ops tend to have a hybrid mix of that and the latest and greatest in civilian cold-weather gear. Often it is mission-dependent, but frequently it comes down to operator preference.

The key thing to pay attention to here is the word *system*. Your clothing is your first line of defense against extreme cold, and you must be able to use a combination of layers to put on when static, or take off when mobile.

The same was true of materials—as footwear in the early days—it was mostly rubber for the overcoat and pants. We'd wear heavy wool pants and cotton long-john type undergarments. I recall when we were some of the first to try out Gortex outer garments. We loved the breathability of them, the lack of bulk, great warmth and wind protection. But they made so much noise when we were trying to walk silently through the dead of a winter night; you could hear this *swish-swish* sound of each man as he swung his arms and legs moving through the forest. To soften this noise, we started wearing our old cotton uniforms on the outside to muffle the loudness.

These days you have a great many choices of materials and styles. The key is to do the research, and then practice with what you purchase to make sure it works for you. Here's a baseline pack list for surviving an extreme cold snap. Again, many of these items should always be on hand, in general.

1. Shelter: If you have to leave your home, pack a portable shelter, such as a tent or emergency bivouac to protect you from the cold.
2. Sleeping bag and blankets: A high-quality sleeping bag and extra blankets can help keep you warm while sleeping.
3. Water and water purification tablets: Pack enough water to last at least three days, and bring water purification tablets in case you run out.
4. Nonperishable food: Pack nonperishable food items such as energy bars, jerky, nuts, and canned goods.
5. Stove and fuel: A small camping stove and fuel can help you prepare warm meals and drinks.
6. First aid kit: Pack a first aid kit with bandages, gauze, and other essentials.
7. Hand and foot warmers: Hand and foot warmers can provide quick sources of heat to help keep you warm.
8. Portable phone charger: A portable phone charger can help you stay connected in case of emergency.
9. Multi-tool: A multi-tool can come in handy for a variety of tasks such as opening cans, cutting rope, and repairing gear.

10. Map and compass: In case you get lost, a map and compass can help you navigate and find your way.
11. Extra batteries: Pack extra batteries for flashlights, head-lamps, and other electronic devices.
12. Personal hygiene items: Pack personal hygiene items such as wet wipes, hand sanitizer, and toilet paper.

Remember to pack these items in a waterproof and sturdy backpack or duffel bag, and keep it easily accessible in case of emergency. It's important to check your supplies periodically and replace any expired or used items.

COLD WEATHER SKILLS TO STUDY

When it comes to dealing with winter survival, there are so many TTPs or Tactics, Techniques, and Procedures, as we call them in special ops. There are way too many to cover in this chapter. But there are some COAs, or courses of action, that are worth mentioning here. The old hands will smile at some of these, and the greenhorns will stand to benefit from learning to avoid hard lessons others have suffered.

Just make sure you feel like you can do these core skills in the cold, wet, wind, and dark to ensure your best chances of success, as practice in the warm and dry is very different. Surviving in cold weather requires a combination of knowledge, skills, and preparedness. Here are some of the best personal cold weather survival skills to have:

1. Building a shelter: Knowing how to build a shelter in cold weather is crucial for survival. A basic shelter can be constructed using natural materials such as branches and leaves.
2. Starting a fire: Knowing how to start a fire in cold weather can provide warmth and a source of light. Practice building fires using different materials and methods, such as flint and steel, or a magnesium fire starter.
3. Finding water: Knowing how to find and purify water is essential for survival in cold weather. Look for sources such as streams, lakes, and snow.

4. Navigation: Knowing how to navigate using a map and compass can help you find your way in cold weather. Practice using a compass and reading a map before you need it.

5. First aid: Knowing basic first aid can be lifesaving in cold weather emergencies. Take a first aid course and pack a first aid kit with essential supplies.

6. Clothing repair: Knowing how to repair damaged clothing or gear can help you stay warm and dry in cold weather. Practice repairing clothing using a sewing kit or duct tape.

7. Self-rescue: Knowing how to self-rescue in cold weather can be crucial in case of emergency. Practice techniques such as self-arrest with an ice axe or using a rope for self-rescue.

8. Winter driving: If you are driving in cold-weather conditions, it's important to know how to drive safely in icy or snowy conditions. Take a winter driving course and pack a winter survival kit in your vehicle.

9. Positive attitude: A positive attitude can help you stay focused and resilient in cold weather emergencies. Practice positive self-talk and visualization techniques.

Remember that prevention is the best defense against cold weather emergencies. Stay informed about weather conditions and avoid unnecessary risks. Being prepared and having the necessary skills can increase your chances of survival in cold weather emergencies.

FOR SELF (Inter- and Intrapersonal)

The following intrapersonal technique is one of the more esoteric things I teach but I've used it to great effect when things were rather dire. I can't explain the full physics of it, so perhaps it is more metaphysical than anything. But somehow, when I'm stuck outside in the freezing cold for long periods, it is inevitable that at some point, you get cold and begin to shiver. When this happens to me, I do two key things:

1. I tell myself, or my brain, that right now, I am not wet; I have

on warm, dry clothing, I have food and water, and sooner or later, I'll get to or make a shelter and fire. This helps to reassure myself and keeps me focused on my realities.

2. I tell myself that the shivering is my body fighting the cold and trying to keep me warm by generating energy to create heat. This is also fact. However, this is where I convince myself to stop fighting the cold and try to embrace it. To accept this temporary and unpleasant experience by allowing it to pass in and through me. To do that, I just breathe in deeply, let the cold enter my lungs, then breathe out slowly and calmly, which always has the effect of relaxing me so that the shivers stop. I'm still cold and in the cold, but somehow, I lessen the misery-fest by embracing it more and fighting it less.

In special ops this technique has a well-known expression called "Embracing the Suck." It sucks, to be sure, but once you embrace it, it somehow sucks less. I leave it to smarter folk than I to explain better.

On the interpersonal level, it's important that everyone be on high alert for not only themselves and how they're responding to the cold, but also, it's imperative that you keep vigilant for the others in your party. Know the signs and symptoms of the impacts of cold on people, and as soon as you see any signs of extreme cold-induced conditions, you must consider that you may need to stop everything and try to resolve or at least reduce the issue.

This may mean stopping to build a shelter, make a fire, get warm and/or dry. It may also mean, quite frankly, getting naked together and sharing a sleeping bag to help prevent hypothermia. This is no time to be squeamish or selective. Their or your life may depend on the ability to overcome the social norms and stigmas about these actions and just get it done—if you want to live.

LOCATION

Obviously, wherever you live, work, or travel will present its own set of challenges, both in terms of weather and what is available for your potential solutions. For those who have lived somewhere cold for a long time already, it doesn't hurt to take a day once a year and review all your cold-weather countermeasure systems and see if there is anything new, different, better, or cheaper to help you refine your winter warfare plans.

For those going somewhere cold, do the research—it could save your life. Learn from locals, they usually know best. And be willing to try new things. But mainly, review your mission, your plans, and your logistics; modify, update, upgrade, or improve wherever you can. At the end of the day, it is a never-ending cycle, and mother nature will test us and she'll always win, unless we are appropriately prepared.

SCIENCE AND SOLUTIONS

Every year there are new technological advances in heating, insulation, building materials, and other warmth-retaining, energy-saving products. The key is to stay current by continually and constantly researching and reviewing. I take advantage of the Google notifications and plug that into mine, so whenever a new tech related to something of interest to me comes out, I get notified, then do the reading and decide whether or not to acquire and implement the technology.

Then there are other things that have been around a while that can also help. Solar panels are good for capturing and storing energy. If you live on a river or stream, you can buy or make a simple water mill; the flow of the current makes the turbine turn and creates energy for you. Often rivers will not freeze unless it gets bitterly cold. And in most cases, in places where it gets cold, they have lots of wind. Windmills are another reasonably affordable way to generate energy that can then be translated into heat by running your stoves, electric furnaces, and small space heaters.

So, whatever your situation, as long as you live, there will be improvements and inventions that can help you. It's up to you to keep up with them to ensure you continually improve your winter readiness and enhance your family's survivability.

To provide thought for what you might be able to do with tech for future indemnification, I'll throw some neat ideas out there. There are several technologies that can be used to protect yourself, your vehicle, and your home from an extreme cold snap. Here are some examples:

1. Personal technologies: There are several personal technologies that can help keep you warm and safe in extremely cold weather. Heated clothing, such as jackets, gloves, and socks, can provide additional warmth. Wearable technologies, such as smartwatches or activity trackers, can monitor your body temperature and alert you if it drops too low. Additionally, portable power banks can be used to charge electronic devices such as cell phones or GPS units.

2. Vehicle technologies: Many modern vehicles have features that can help protect you during an extreme cold snap. Remote start systems allow you to start your vehicle from inside your home, so it can warm up before you get in. Traction control systems can help prevent your vehicle from slipping on icy roads. Additionally, some vehicles are equipped with heated seats, mirrors, and windshield wipers.

3. Home technologies: Several home technologies can help protect you from an extreme cold snap. Programmable thermostats can automatically adjust the temperature of your home to ensure it stays warm. Smart home technologies, such as smart plugs and smart lights, can be used to turn on appliances and lights from a remote location, ensuring your home stays warm and well-lit even if you're not there. Insulated windows, doors, and walls can also help keep cold air out and warm air in.

4. Backup generators: Installing a backup generator can help keep your home powered in the event of a power outage during an extreme cold snap. A generator can power essential appliances such as heaters, refrigerators, and lights, ensuring your family stays warm and comfortable.

By utilizing these technologies, you can help protect yourself, your vehicle, and your home during an extreme cold snap. It's important to always have a backup plan and to be prepared for emergencies, even if you have access to these technologies.

Bottom line, stay warm, stay dry, practice, learn, improve, repeat.

4

A Hazy Shade of Winter

SNOW—THE SITUATION

We move on to a phenomenon consisting of grains, flakes, and pellets, though what lies within each of those manifestations will always be a nucleus of dust or a similar particle surrounded by frozen water. What we are talking about here is snow in all its forms. For most of us, falling snow is greeted with a sense of joy, and even with a degree of childish happiness, but it also has a dark, perilous side. So, let's plow on and become wise to the threats that exist when the white stuff comes-a-calling.

Did I say white? Well, actually snow isn't white at all, it's translucent—only appearing white to the eye when light passes irregularly through it. The many sides of the ice crystals cause diffuse reflection of the light spectrum, which results in snowflakes, grains, and pellets appearing to be white in color, both in the air and on the ground.

Meanwhile, thinking of those translucent flakes, believe it or not, no two are exactly the same; every single one of them is original and unique. The tiniest variations of air temperature, humidity, and wind make for that complete individuality. I can't say the same for snow pellets or snow grains, but in terms of the flakes they are a colossal army made up of millions or billions of individual soldiers.

There are many types of snow; the Inuit people of the high Arctic have more than fifty names for it. For example, the dialect spoken in Nunavik, Canada, includes *pukak* to refer to crystal-like snow that appears to be salt-like; *matsaaruti*, meaning wet snow/sleet; and *qanik*, which simply refers to falling snow. But in any case, and whatever the name, I guess most of us know what it takes to build a resolute snowman!

Before we get into the teeth of things, and always worth pointing out and repeating to any doubters you may come across, global warming does not necessarily mean we will see *less* falling snow in the future. Warmer temperatures translate into a *greater* capacity for higher amounts of precipitation being held in the atmosphere, and when it's cold enough, that also means exaggerated and potentially debilitating snow events.

In addition, climate change can also create added complexities in and around the Arctic and Antarctic, where any sudden warming of the stratosphere above the poles could well result in the ordinarily stable polar vortexes becoming weak and dislocating, cascading freezing cold air and high volumes of snow into populated land masses farther south and north, depending upon the hemisphere. We might well hear about much more of this chaotic and unpredictable big snow events as the years unfold.

What I'm not going to be mentioning here, in terms of the threats that snow can bring, is hypothermia. We covered that particular peril already, though it still goes hand in hand with the other perils I'm about to delve into. Akin to every other weather-related hazard, it's entirely appropriate to be respectful of snow and understand it, but not necessarily to fear it. If you do happen to have a fear of snow, perhaps brought on by a bad slip, a traffic incident, or maybe being walloped by a snowball in the face, then you are suffering from what's called Chionophobia, *chion* being the Greek word for snow. So, what are those other perils worthy of your attention? Well, the main threats from snow with accompanying ice include snow blindness, accidents or incidents of numerous kinds, and the daddy of them all, avalanches.

Let's take snow blindness first or, to give it its scientific name, photokeratitis. This painful condition is due to the overexposure of your eyes to the intense ultraviolet light of the sun, whether by direct entry into the eyes or indirectly by bouncing up off lying snow. This is why those higher-risk groups partaking in snow sports, or traversing major snowfields, should always take appropriate precautions.

There is another form of snow blindness worthy of mention at this point, which is somewhat akin to perilous sand and dust storms, in terms of impairment or loss of vision. When the wind is blowing and the snow is falling, blizzards can result, which in turn reduce visibilities to a point where it is genuinely difficult to see your hands in front of your face. Given that, perhaps with an added element of darkness and any lack of topographical reference points, acute disorientation can suddenly become a real issue. Any explorer or mountain climber will testify to how this form of blindness can be perilous, though you need not be conquering Antarctica or climbing Mount Everest to fall foul of this peril. In the city of Buffalo, New York, in December 2022, a blinding Christmas blizzard of epic proportions led to people becoming lost, even within a few yards of their own homes, which sadly contributed to a considerable loss of life. There are numerous other examples of such events and there will be many more to come.

Turning now to accidents or incidents. We've all been, haven't we? Snow lying deep, crisp, and even; ice in patches or stretches; and you and I playfully frolicking or attempting to gingerly negotiate what's under our feet. Frozen precipitation in all its forms presents numerous hazards to both lives and limbs, just ask any doctor or nurse at your local hospital during the onset of a wintry spell. Yes, some of us are more accustomed to it and able to deal with the various hazards more easily than others. Some of us actually put ourselves directly in the firing line for falls and other similar accidents, which is why insurance companies charge winter sports enthusiasts and mountain climbers far higher premiums. In simple terms, ice or snow underfoot creates a stability problem, with freezing rain (rain freezing upon impact with frozen surfaces) the element that arguably causes the greatest distress, particularly affecting all forms of transport.

Numerous as they usually are when the ground is carpeted and also icy, it's not only slips or falls that we are concerned about here. What about that snowball containing a shard of glass or a small rock? What about the speeding motorist who hits a patch of snow or ice and loses control? What about the terrible train derailment in a blizzard, or the sports roof that collapses on people sheltering from a snowstorm? And then there are those power lines that fall and leave thousands without electricity. The guy who dies of a heart attack by overexerting himself while clearing snow from his path, not to mention the countless thousands stranded at airports and service stations. The 1987 film *Planes, Trains, and Automobiles* suddenly springs to mind!

I could go on and on about the types and degrees of frozen precipitation incidents, but suffice to say that in the verification work I do for the legal fraternity, frozen precipitation cases are by far the most numerous, and more often than not the most marginal in terms of the outcomes. Marginal? Well, there are often very fine lines when it comes to faults, claims, and outcomes concerning the wintry elements; for example, did the roof cave in due to the weight of snow, or was it simply down to a dodgy roof? Was the driver blinded by a sudden onset of a snow shower at the time of the crash, or was it just a case of careless driving? And was that body in the woods subjected to snow and freezing temperatures that delayed decomposition, or had the body only been there for the past day or so?

I'm going to finish my bit on snow-driven impacts by concentrating on something that is truly manifest and can be very deadly, namely avalanches. An avalanche is a mobile mass or wall of snow, often intermixed with rocks, soil, and debris, that cascades down a hill or mountain at high speed.

They can be set in motion by wind, rain, warming temperatures, fresh snow, and earthquakes or tremors. The catalysts can also be man-made, triggered by skiers, hikers, snowmobiles, and vibrations from machinery or construction—so let's call them what they are, human disrupters.

The four types of avalanches are:

1. Loose snow: Often occurring on steep slopes, more often than not following a heavy snowfall of several inches.
2. Slab: In essence, a sliding plate or block of ice, like a snow-covered ice cube on the move.
3. Powder snow: A turbulent flow of snow grains mostly suspended in the air, often associated with unstable gusty winds.
4. Wet snow: This particular one concerns fallen snow being undermined or melted by rainfall or warmer temperatures/direct sunshine. The spring season provides the optimum time for this type of event.

Dependent upon the steepness of the hill or mountain and the size and weight of the cascading snowpack, avalanches can reach speeds of up to 200 mph, though the average speed is said to range between 70 and 100 mph. Either way, they are not to be outskied or outrun.

In terms of the peril to life, by far the major cause of death from avalanches comes from asphyxiation—being buried and losing the ability to breathe. The secondary cause of death is by way of blunt trauma, essentially succumbing to being hit by speeding snow, ice, rocks, trees, or debris, usually impacting the head or neck. View this as similar to being hit by a train moving at full speed. There is a third, a slower death due to being buried in a tomb of deep snow, not being found in time, and subsequently succumbing to the word I said I wouldn't mention here but have now, namely hyperthermia.

If we are looking for the main avalanche countries or regions, then Switzerland, Afghanistan, and Nepal appear to be at the top of the list, while Colorado wears the dubious crown as the US's avalanche death capital. But take it from me, virtually anywhere in or around mountains or hills with a sufficient snow covering can be at risk, with the inerrant instability of the spring months being the time of year to be most wary, given the added ingredient of global warming.

On average across the world, less than two hundred people die from this phenomenon per year, but throughout history there have been many cases of terrible avalanches that have taken numerous lives. The worst case occurred in May 1970, when the Ancash earthquake, located in the Pacific Ocean just off Peru, resulted in a calamitous Huascarán snow, ice, mud, and debris avalanche, which accounted for an estimated 30,000 deaths.

So, there we have it, and akin to many elements of the weather, that pretty, flaky stuff most of us love to dream about, whether in the form of snowstorms, compacted snow/ice, or avalanches, can have a very deadly underside. Over to Mykel for his escape and survival plan!

SNOW—THE SURVIVAL

I spent my first few years in special forces on a winter warfare team. This was back during the Cold War of the eighties, when our primary mission was to be dropped behind Russian lines and cause as much havoc as possible through sabotage, ambush, snipers, and small backpack nukes—all the while living off the land and trying to win the hearts and minds of Russian peasants who might be willing to wage a guerrilla war against the tyranny and oppression of their authoritarian communist overlords.

So, we did a lot of training in the snow. Alaska, New Hampshire, and Norway were the norms. I loved it all, from snowshoes to igloos, ahkios, skis, and all the mountain-climbing kit you could ever want. And they gave me three wall lockers full of cold-weather clothing gear to keep me warm. But then they sent me on one mission to Honduras, where my only issue of clothing was swim trunks and sandals. I changed units after that, and never went back. That's why I live in Florida—it's the closest thing we have to jungle on the mainland. But let's look at ways and strategies for dealing with all the variations of the dirty word which I spell as S.N.O.W.

Jim did a great job of breaking down snow into the different kinds, types, issues, and concerns. Here, I'm going to address the varied types and their immediate and direct impacts on you as a human.

On foot:
- Slips, trips, and falls: Snow and ice can make walking surfaces slippery, leading to accidents.
- Frostbite and hypothermia: Exposure to cold temperatures for an extended period can lead to frostbite and hypothermia.
- Poor visibility: Heavy snowfall can reduce visibility, making it challenging to see and navigate surroundings.

In a car:
- Reduced traction and control: Snow and ice can reduce the traction and control of the vehicle, increasing the risk of accidents.
- Blocked roads and highways: Heavy snowfall can lead to blocked roads and highways, causing traffic congestion and delays.
- Danger from falling ice: Falling icicles or chunks of ice can pose a danger to vehicles and their passengers.

At home:
- Difficulty accessing the property: Snow and ice can make it difficult to access the home, particularly for those with mobility issues.
- Frozen pipes: Cold temperatures can cause pipes to freeze and burst, causing damage to the home.
- Power outages: Snow and ice storms can knock out power, leaving people without heat or electricity.

Overall, snow and ice are impactful, and it is essential to take appropriate precautions to stay safe and avoid accidents. Now that we've seen the overview, let's break it down.

TACTICAL

Let's look at some of the personal measures you can take or consider as you prep yourself for facing snow. If you are caught out in a snowstorm far from anywhere, it is crucial to take immediate steps to protect yourself and increase your chances of survival. Here are some things you can do:

1. Find shelter: Look for shelter to protect yourself from the wind and snow. If you can't find a building or other structure, create your shelter using materials such as tree branches, leaves, and snow blocks.
2. Stay warm: It's essential to stay warm to avoid hypothermia. If you have a heat source, use it. If not, use layers of clothing and blankets to stay warm.
3. Stay hydrated: Drink plenty of fluids, including water and hot liquids, like tea or soup, to avoid dehydration.
4. Avoid overexertion: Try to conserve energy and avoid over-exertion, which can cause sweating and increase heat loss.
5. Signal for help: If you have a signal device, such as a whistle or flare, use it to signal for help. You can also create a distress signal using rocks or sticks in the snow.
6. Keep yourself fed: If you have any food, eat it to keep your energy levels up.
7. Stay calm and stay put: Don't try to walk or travel in a snowstorm. Stay put and wait for rescue. Moving around in a snowstorm can be disorienting and make it challenging for rescuers to find you.

It's always essential to be prepared for emergencies, including having appropriate clothing, equipment, and provisions if you plan to be out in remote areas. However, if you find yourself caught out in a snowstorm, taking these steps can help increase your chances of survival. What you have or don't is often the main factor in what will save you, unless you have mad skills or great luck. Preparation is key and we teach throughout how to build kits for every potentiality.

OPERATIONAL

If you have a vehicle, the same list of supplies for cold weather applies here (snow and ice removal tools, sand or cat litter, jumper cables, blankets, extra clothing, first aid kit, etc.)

It's always essential to be prepared for emergencies, especially when driving in areas that experience snowstorms. Having your vehicle supplied, equipped, and properly maintained can help you stay safe and comfortable in case of an emergency.

STRATEGIC

The home is the best place to be for any snowstorm. If well prepared, it's one of the few storms you can often make snuggly and weather without too much fear of loss of life or property, unless you have no heat, or live where avalanches, boulders, or trees can take you out. But in general, folks who live in these regions have learned how best to adjust, so learn from them. If you're somewhere that it may be unusual but possible to get a snowstorm, here's a good start for home prep. If you live in an area that experiences long, dangerous snowstorm blizzards, it's important to be prepared for potential power outages and disruptions to transportation and communication systems. Again, the same list of supplies to have on hand for cold weather applies here. (Nonperishable food, water, battery-powered items, warm clothing, first aid kit, etc.)

If you don't have a fire pit or fireplace in your home and you're looking for a temporary solution, there are still several options available that can provide warmth and a cozy atmosphere. Here are some ideas:

1. Portable fire pit: A portable fire pit is a great option for those who don't have a permanent fire pit or fireplace. They are easy to set up and come in a variety of sizes and styles.
2. Tabletop fire bowl: A tabletop fire bowl is a smaller, more compact version of a fire pit that can be used on a tabletop or other flat surface.
3. Indoor electric fireplace: An indoor electric fireplace is a

great option for those who don't have a traditional fireplace. They are easy to install, and some models even come with built-in heaters.

4. Candles: Candles can provide a warm, flickering light, and a cozy atmosphere. Use a variety of sizes and scents to create a relaxing environment.

5. String lights: String lights can be used to create a warm and inviting atmosphere and can be hung around a room or outside.

6. Oil lamps: Oil lamps are a classic way to add a warm glow to a room and can be found in a variety of styles and sizes.

7. DIY fire pit: If you have a backyard or outdoor space, you can create a DIY fire pit using materials such as bricks, stones, or cinder blocks.

These options can provide a cozy atmosphere and warmth in your home, without requiring major renovations or permanent installations. We have a fire pit, but are too far south to have a fireplace, so I bought a nice tent stove that looks like a mini wood stove with all its functions and piping to run out of a window. We also have the fireproof ground mat and window portal for the stovepipe. It means we can have a safe fire inside our home or anywhere we go. They're more work to chop wood smaller, but the wood lasts much longer with reduced air flow, and it burns much hotter for faster cooking and boiling. We also have one of those gas fires that you can sit around inside the pool cage in the winter, and they provide heat in emergencies, too. Just keep them outdoors.

MY STORIES

I've spent more time than I care to admit in the cold, and the things that fascinate me most are the things I learned and did that I didn't think were possible, like making fire from ice, using polished convex pieces of ice on dry, sunny, freezing cold days, with good tinder and making a fire. It gave me great confidence and I recommend it if you ever have the time for some fun play.

I spent many cold winters as a kid with no water, no heat, no power, and sometimes, no doors or windows, and an outhouse. As a result, I was never a fan of snow. But one particular winter as a teen, I remember getting into an uncontrollable shiver. Then, with nowhere else to go, no options to turn to and no help from anyone, I kind of made a peace with freezing to death, and I breathed in deeply and just let go. I stopped shivering! Since then, I have used that same technique successfully for both self-control and falling asleep.

For a long time, I had a white wolf, also known as an Arctic wolf. It's a long story, the kind best told over a drink and "round a fire," but he was no dog. He was more like a son. Lukos was his name, and it just meant wolf in Greek, which I'm nearly half. But one day, we were in a bad blizzard in Minnesota, the minus 40-degree kind where your eyeballs hurt. My sons and I were hunkering down, but Lukos was at home in the storm. He sang the entire blizzard, and my lads and I worried not; he soothed us.

FRIENDS AND NEIGHBORS

Best practices for your friend network don't vary much here. But let's look at several things you can do to help your friends during a blizzard. Here are some ideas:

1. Check in on them: If you know someone is in the area affected by the blizzard, reach out and check on them. Offer to help them with anything they need, such as bringing them groceries or running errands. Make sure they're safe and have everything they need.

2. Offer to share supplies: If you have extra food, water, or blankets, offer to share them with your friends. This can be especially helpful if they're unable to leave their home due to the blizzard.

3. Help shovel snow: If you're physically able, offer to help your friends shovel snow from their driveway or sidewalk; this can be especially vital for elderly or disabled neighbors who may not be able to do it themselves.

4. Share snow removal equipment: If you have snow removal equipment like a snowblower or snow shovel, offer to share it with your neighbors who may not have their own.

5. Provide transportation: If you have a vehicle that's capable of driving in the snow, offer to provide transportation for your friends if they need to go somewhere. This could be to a nearby store or to a warm shelter if their home loses power.

6. Stay connected: Check in with your friends regularly to make sure they're doing okay. This can help to alleviate feelings of isolation and can provide emotional support during a difficult time.

Overall, it's important to be a good friend and neighbor during a snowstorm. Small acts of kindness can go a long way in helping them feel supported and cared for during a difficult time.

HOME AND PROPERTY

If you're in a place where snowstorms are a real threat, you should consider investing in structural as well as superficial preparation. If you live in an area that is susceptible to blizzards, it's important to take steps to prepare your home for the worst. For this, I refer back to our section on cold, which details home and yard maintenance and materials that enhance preparation for extremes.

In addition, specifically for snow and ice:

1. Install snow fences: Install snow fences to prevent snowdrifts from accumulating in certain areas of your yard.

2. Cover outdoor furniture: Cover outdoor furniture with tarps or move it inside to prevent damage from snow and ice.

3. Install heated driveways and walkways: Install heated driveways and walkways to prevent snow and ice from accumulating.

4. Shovel snow promptly: Shovel snow promptly to prevent it from accumulating and causing damage to plants or outdoor structures.

By taking these steps, you can help protect your yard from snow and ice damage and ensure that it remains in good condition throughout the winter.

VEHICLE

If you have to get out and drive in snow and ice, there are some good measures to consider. Driving on snow and ice can be challenging and requires special precautions to ensure safety. Here are some tips for driving on snow and ice:

1. Slow down: Reduce your speed and give yourself more time to react to changes in road conditions.
2. Increase following distance: Increase the following distance between your car and the car in front of you to give yourself more time to brake.
3. Avoid sudden braking and acceleration: Sudden braking or acceleration can cause your car to slide on icy or snowy roads, so be gentle with the gas and brake pedals.
4. Use winter tires: Winter tires provide better traction on snow and ice, so consider investing in a set of winter tires.
5. Use low gears: Use low gears when driving on snowy or icy roads to reduce the risk of skidding.
6. Use antilock brakes (ABS): If your car is equipped with ABS, use it to help prevent skidding and maintain control of your vehicle.
7. Use your headlights: Turn on your headlights to improve your visibility and make it easier for other drivers to see you.
8. Keep your car well maintained: Regular maintenance of your car, including brakes, tires, and windshield wipers, can help ensure that it is in good condition for winter driving.

Remember that driving on snow and ice requires extra caution, and it's essential to adjust your driving habits to the road conditions. By following these tips, you can increase your chances of staying safe on snowy and icy roads.

For first aid and applications, much of the same applies for the cold; a list can be found in Appendix II.

MOBILITY PLANNING

Sometimes, you just have to get somewhere, even in dangerous snow and ice conditions. If you do, all the other chapters apply, but for ice, it's a bit different for driving. If you find yourself in a dangerous skid on ice while driving a car, it's essential to stay calm and take the appropriate steps to regain control of your vehicle. Here are some steps you can take:

1. Steer in the direction of the skid: If your car is sliding to the right, turn your steering wheel to the right. If your car is sliding to the left, turn your steering wheel to the left. This will help you regain control of your vehicle.
2. Do not overcorrect: Do not overcorrect your steering or overreact by jerking the steering wheel in the opposite direction, as this can make the skid worse.
3. Do not brake suddenly: Avoid sudden braking, as this can cause your car to skid even more. Instead, gently apply the brakes and release them as necessary.
4. Stay off the gas pedal: Do not accelerate while in a skid, as this can make the skid worse. Instead, take your foot off the gas pedal and allow your car to slow down naturally.
5. Be patient: Skids can take several seconds to resolve, so be patient and wait for your car to regain traction.
6. Focus on where you want to go: Look where you want to go, and steer in that direction. Your hands will follow your eyes.
7. Stay aware of your surroundings: Keep an eye on other cars and obstacles on the road, and be prepared to take evasive action if necessary.

Remember, prevention is the best way to avoid skids on icy roads. Drive at a safe speed, keep a safe distance between your car and other vehicles, and avoid sudden braking or acceleration. If you do find yourself in a skid, stay calm, and follow these steps to regain control of your vehicle.

STATIC PLANNING

I think most folk who live with seasons know what to do at home to get ready for winter and we've already covered what to do temporarily and permanently to blizzard-proof, too. Something that was not covered or included is the human fatigue factor.

All parents know to have movies, music, books, and board games for long stints of isolation to fight off boredom and provide positive mental distraction. But sometimes, you just get caught with nothing. These games can be applied to any chapter of this book. I happen to love these! There are many fun games that families can play together without needing any special equipment or board games. Here are some of the best games families can play when they don't have others:

1. Charades: Charades is a classic game that only requires your imagination. One person acts out a word or phrase without speaking, and the others try to guess what it is.

2. Twenty Questions: One person thinks of an object, and the others take turns asking yes-or-no questions to try to guess what it is within twenty questions.

3. Categories: One person chooses a category (such as food, animals, or colors), and everyone takes turns naming something that falls within that category.

4. Pictionary: One person draws a picture, and the others try to guess what it is.

5. I Spy: One person says "I spy with my little eye" and gives a clue about something they see, and the others try to guess what it is.

6. Word Chain: One person says a word, and the others take turns saying a word that starts with the last letter of the previous word.

7. Name That Tune: Play (or hum) a few seconds of a song and see who can guess the name of the song and the artist.

8. Two Truths and a Lie: Each person takes turns sharing three statements about themselves, two of which are true and one is a lie. The others try to guess which statement is the lie.

These games are all easy to play and can provide hours of entertainment for the whole family.

PERSONAL

When I am in a snow environment, I am very aware that it can kill me. I always joke and say, for problems in scuba diving, the solution is always to go back up; for skydiving, it's only ever the reserve chute. And the same for the cold. I say, in the jungle, everything can kill and eat me, but also, I can kill and eat everything, too. In the cold, it's hard to find food to eat, and the cold is only one thing, but it is a constant and it can kill you any moment you are not ready. So, make sure you have a checklist that you don't leave home without when dealing in a cold weather environment. When venturing out in extreme snow blizzards, it's important to keep essential items on your person in case of an emergency. In addition to what's already been discussed for the cold, such as navigation and communication tools, food and water, first aid, etc., wear waterproof gear, including a waterproof jacket and pants, to stay dry and avoid hypothermia.

I am a gadget person who realizes personal limitations, so I'm always looking for an advantage or an edge against mother nature so, as with the cold-weather tech and gear recommendations, I lean into heated clothing items (hand warmers and the like), portable heaters, and other items and gear of convenience with built-in heat elements.

1. Heated jackets: Heated jackets use rechargeable batteries to provide heat through built-in heating elements.
2. Heated vests: Same idea as jackets but can allow greater flexibility of movement.
3. Heated blankets: These can also be plugged into and used in a vehicle.

4. Heated pants and pants liners.
5. Heated socks and insoles.
6. Heated boots: Some of these are wearable indoors/outdoors.

These gadgets can provide extra warmth in cold weather and can make being outside or inside more comfortable. However, it's important to use these gadgets safely and follow all manufacturer's instructions to avoid accidents or injuries. I own lots of these, and I live in Florida!

INTRAPERSONAL

I'm not a fan of cold, but interestingly enough, I do transition back into it a lot more readily than I'd like to admit. I hate it at first, but then I accept and get right into the brisk and invigorating aspects of it. I find there are things I can do in the immediate, but also, in the bigger picture to make me more mentally ready for operating in the cold. Since being extremely cold can be challenging, it's important to take care of your mental health as well as your physical health. Here are a few mental ways of dealing with being extremely cold:

1. Stay positive: Maintaining a positive attitude can help you stay motivated and resilient in the face of cold temperatures.
2. Focus on your breathing: Focusing on your breath can help calm your mind and body, and regulate your body temperature.
3. Engage in physical activity: Exercise can help warm you up and release endorphins, which can improve your mood and overall well-being.
4. Stay connected: Stay connected with loved ones, friends, or community members to combat feelings of loneliness or isolation.
5. Practice relaxation techniques: Relaxation techniques such as meditation, yoga, or deep breathing exercises can help reduce stress and anxiety.
6. Engage in hobbies or activities you enjoy: Doing something

you enjoy can help distract you from the cold and improve your overall mood.

7. Seek professional help: If you are experiencing extreme cold-related anxiety or depression, seek professional help from a mental health professional. I say this as long bouts of gray skies and insular emotions from being trapped inside can have a powerfully negative impact on folks. Even the mountain men, as hardy a bunch as there ever was, knew all too well the madness of cabin fever. I have a master's in psychology, and cabin fever is as real as anything if you get it. So try to be aware of it, and do these things to hopefully prevent it. Prevention is always best.

It's important to remember that everyone responds to cold temperatures differently, and it's okay to seek help if you are struggling. By taking care of your mental health and well-being, you can better cope with extreme cold temperatures and enjoy the winter season.

SECTION III
WATER AND STORMS

A storm in a teacup is the smallest storm you are ever likely to face; all other storms come in tankards and trucks.

5

Greased Lightning

THUNDERSTORMS—THE SITUATION

I now turn to perhaps my most favorite weather of all (and that takes some doing, given a plethora of choices): thunderstorms!

Ever since I was a young boy, staring out from the large bay window in my parents' home near Manchester, England, at an ominous and threatening black sky illuminated by occasional bolts of lightning, thunderstorms have fascinated me to my absolute core. Maybe it's the same for you—after all, these beasts of the heavens are majestic, beautiful, ultra-powerful, and ultimately very frightening, with a primeval, ethereal quality that often lands them star roles in many horror films. Worshipped and feared in equal measure during prehistoric times, and through every age of man since, they are both life-giving and a life-taking, and you really do need to avoid getting yourself in the category of the latter.

When we talk about the threat that thunderstorms pose, then heavy rain leading to flash floods and sudden gusty winds that are capable of transporting debris are two of the main threats, along with tornado activity. However, in this chapter Mykel and I will focus on two other deadly items that may threaten your well-being or your life: lightning and hailstones.

In my occasional work as a forensic meteorologist, it has been my sad but dutiful task to report upon and explain to coroners and investigating police how and why the victims of lightning strikes ended up as just that, victims. You see, when thunderstorms appear, there are ways and means of putting oneself in the right place or the wrong place, and it is my job to differentiate and then explain why choosing the wrong place was a huge mistake. Of course, it is possible to be damn unlucky and become a victim even in what might be deemed as a safe and appropriate place to be, but lightning can strike out of the blue from up to ten to fifteen miles distance and in the most unlikely of places. Putting it simply, there are no free passes when it comes to your exposure to lightning strikes, but it is an odds game; you can reduce your odds of being struck if you are situated in safer zones and can raise your odds of being struck if you place yourself in areas of greatest exposure. Mykel will advise more on the places to be and the places to avoid, but at this point, just know that outside open spaces are your enemy when thunderstorms are prowling. Also, if you happen to be one of the tallest features on the land or seascape, then you, my friend, are at greater risk. Allow me to explain.

Thunderstorms only occur within cumulonimbus clouds—big, inky black, with an anvil top, at least when viewed from a distance. Lightning forms when ice crystals and water droplets within these clouds collide, creating positive and negative charges. The charges then separate due to convective forces, and once far enough apart, the law of attraction comes into play. By joining the negative dashes and positive crosses, lightning can then discharge.

Cloud-to-ground strikes occur when a channel of negative charges, called a stepped leader, zigzags downward, heading for the opposite positive charges on the ground surface. That could be a tree, a building, a fence, a flagpole, a boat, a cow . . . or it could even be you! You, and all the objects I mention, are usually emitting positive leader charges, and when the two opposite charges meet, we have an electrical current! It all happens within milliseconds and is invisible to the eye, so once selected by the stepped leader, there is no escape.

While we are talking positive and negative charges, it's useful to know that the very tops of cumulonimbus clouds can release massive positive charges that seek out negative charges, sometimes found just below the earth's surface, resulting in damaging super strikes, which account for around 5 percent of all lightning strikes. Physics can sometimes be complicated and very deadly!

To whet your appetite a little more, allow me to provide you with some general facts on thunderstorms and lightning:

Lightning can heat the air, or objects it passes through, to some fifty thousand degrees Fahrenheit, which is five times hotter than the surface of the sun. There are between 15 to 17 million thunderstorms worldwide every year, and at any given time there are some two thousand thunderstorms in motion, with tropical areas being their favorite haunting grounds. If we are seeking the country harvesting the most lightning strikes per year, then look no further than Brazil, while the highest lightning density country award goes to Singapore. However, and according to NASA Earth Observatory, Lake Maracaibo in northern Venezuela boasts the highest flash extent density anywhere in the world. The region's topography fuels weather patterns that make it a lightning-production factory. Meanwhile, some 25 to 30 million lightning strikes are recorded each year, mostly during the summer months, but not exclusively so. If you are genuinely fearful of thunder and lightning, then you are a member of the keraunophobia club!

So now you are primed a little more, and whether you are keraunophobic or not, there are five types of lightning strikes you should be aware of:

1. Direct strikes: Not the most common, but certainly the deadliest of all strikes, with the victim becoming part of the main discharge channel. Open and exposed spaces, such as fields, golf courses, waterways, ridges of hills, mountains, and even on the tops of buildings, act as the perfect stage. Once struck, the current entering the body will move potently through one's cardiovascular or nervous system, while what's known as flashover skims and burns the skin's surface. Either way or both, it's not pretty!

2. Side strike: The meaning is written on the can. The main stepped leader strike shakes hands with a taller nearby object such as a tree, while an offshoot side current jumps across and into the victim, who might be standing only a few feet away from the object, more often than not sheltering from rain or hail.

3. Ground current: This one is responsible for the most deaths or injuries, given it impacts large surface areas. A strike hits a tree, a portion of field or a concrete floor, for example, and the current then travels just under the ground's surface, sending high voltage surges upward into anything or

anybody on that surface. This is a classic for the deaths of many cattle at the same time, while those playing outdoor sports such as football or baseball also need to be aware of how this one might pan out.

4. Conduction: We've all done the science in class, haven't we? Heat, and in this case electricity, travel happily through metal wires, metal fencing, metal rods, and the like— though lightning is *not* attracted to metal per se, which is a common fallacy. But, if you happen to be touching metal when there is a lightning strike, even some distance away, you could become part of the current as it travels within the channel. This one is a big indoor risk, particularly involving the plumbing conduits within bathrooms and kitchens, or electrical circuits attached to computers, headsets, wired telephone systems, hair dryers, shavers, and the like.

5. Streamers: These are offshoots of the main, positively charged, upward streamer. As the main streamer makes contact with the main downward step leader, several other upwards streamers from the ground, or what's on the ground (you), make contact with the negative charged cloud and result in separate, but equally deadly, discharges.

It all sounds quite disturbing, particularly knowing that you can play a big part in facilitating the eventual development of a lightning strike with your invisible upward streamer coming out of the top of your head. However, populations are far more aware of the dangers than in decades past, and global fatalities are, thankfully, fewer than previously. The US, for example, averages close to twenty-five lightning-strike deaths per annum, while the UK averages below five. Far higher death rates can be found in places such as Brazil, South Africa, Colombia, Indonesia, India, Malaysia, and Thailand, to name but a few higher death toll countries, culminating in an estimated annual average of twenty-five thousand fatalities and hundreds of thousands of injuries, which is plenty too many! However, and to settle your nerves a little, be aware that (and unless you purposely put yourself in a prone place on purpose) modeled statistics suggest you have about a 1 in 15,000 probability of being struck by lightning, rising to a 1 in 5,000 risk in those countries mentioned above.

Mykel will enlighten you of the signs to be aware of if you suddenly find yourself within a growing electrical field, and what you then might do, but I'm going to finish my piece with a quick look at another thunderstorm threat, namely hailstones. Make no mistake, these hard nuts of the sky do present a clear and present danger to life and limb if they are big enough, and very often they are.

Hailstones vary in size, and most won't be big or weighty enough to cause injury, let alone death—those of say 5–10 mm in diameter, that is. Painful maybe, but that's about it. However, it's the larger and heavier ones in excess of 20 mm and upwards of 70 mm diameter that are of obvious concern, given their weight and speed of descent. These stones of pure ice are likely to be travelling at anywhere between 10 and 40 mph, with the larger ones falling at the greater speeds. For comparison, a golf ball has a diameter of 42.7 mm. Can you imagine a sky full of golf balls, or even bigger and harder balls, potentially raining down on top of your head?

To place the largest hailstones into some sort of danger perspective, back in April 1888, and according to the World Meteorological Organization, the worst-ever death toll from hailstones occurred in Moradabad, Uttar Pradesh, India, with hailstones the size of "goose eggs, oranges, and cricket balls." However, deaths by hailstones are a rare event, with less than a dozen fatalities recorded across the US and UK since the year 2000. On the other hand, property and crop damage are entirely different matters and much more manifest—just ask your friendly property or agricultural insurer. So, surely the lesson here is not to find oneself in a shoddy wooden shed in the middle of a farmer's field during the most severe thunderstorm of the year!

Over to Mykel for his take on how to avoid being electrified or cracked on the head by a big ball of ice!

THUNDERSTORMS—THE SURVIVAL

For this chapter, we're going to drill down on two of the most dangerous aspects of storms which, in this case, are lightning and hail.

FACTS AND SCIENCE
Here are some basic scientific facts about lightning:

1. Lightning is an electrical discharge that occurs between the atmosphere and the ground, or within a cloud.

2. Lightning is caused by the build-up of electric charges within a thunderstorm cloud. These charges are created by the collision of ice particles and water droplets within the cloud.

3. Lightning can strike the ground or objects on the ground, such as trees, buildings, or people.

4. Lightning can be dangerous and even deadly, as it can cause fires, electrical surges, and other damage.

5. Thunder is the sound produced by lightning as it heats up the air around it, causing it to expand rapidly.

6. Lightning can occur during any time of the year, but is most common in the summer months.

7. Lightning strikes are more common in open areas, such as fields or golf courses, than in densely populated areas.

8. Lightning can be detected and tracked using weather radar and other instruments.

9. Lightning can also occur within a cloud, between clouds, or from a cloud to the atmosphere, without necessarily striking the ground.

10. Lightning is a powerful force of nature and has fascinated scientists and people for centuries.

TACTICAL

For this section, we're just going to focus on personal measures you can take. Let's start with a good rule of thumb for determining the immediacy of the threat to you directly. We do this by estimating distance: the general five-is-one rule is, for every five seconds after you see the flash of light, the storm is approximately one mile from your position. So, you can estimate the distance of lightning by counting the number of seconds between seeing the lightning and hearing the thunder. Since light travels faster than sound, you will typically see the lightning before hearing the thunder.

Similarly, you can use the flash-to-bang method, which involves counting the number of seconds between seeing the lightning flash and hearing the thunderclap. For every five seconds you count, the lightning is approximately one mile (or 1.6 kilometers) away.

For example, if you see a lightning flash and then count to ten before hearing the thunder, the lightning is approximately two miles (or 3.2 kilometers) away. Keep in mind that this is just an estimation, and lightning can strike outside the five-mile radius. It's important to take precautions when you hear thunder, as lightning can still be dangerous even if it seems far away.

So, what is the safe zone? Technically, there isn't one, but the general thinking is that five miles is usually safe. If you don't count twenty-five seconds from flash to bang, you're in the danger zone. What can you do if you get caught out?

Lightning can strike a person or object even if it is several miles away from the actual thunderstorm. This is because lightning can travel horizontally for several miles before it strikes the ground or an object.

However, the likelihood of being struck by lightning decreases as you move farther away from the storm. On average, lightning strikes occur within a radius of about five miles (or eight kilometers) from the center of the thunderstorm.

It's important to take precautions during a thunderstorm and avoid being outside, as lightning can strike without warning and can be very dangerous. Seek shelter in a sturdy building or vehicle, and avoid standing near tall objects or bodies of water. If you are caught outdoors with no shelter available, crouch low on the ground with your hands on your knees and avoid lying flat.

OPERATIONAL

For this section, we're going to assume you're using your car/boat/plane to get to safety, and as such, you may have fam, friends, and pets with you, so we want to cover some techniques you can consider for safety while traveling.

If you are in a car during a bad lightning storm, there are several things you can do to protect your family:

1. Stay inside the car: If possible, stay inside the car and avoid getting out until the storm has passed. The metal frame

of the car will provide some protection against lightning strikes.

2. Close the windows.

3. Avoid contact with metal: Avoid touching any metal parts of the car, including door handles and things such as radio dials. This will help to prevent any electric shock in case of a lightning strike.

4. Avoid parking under trees: If you must pull over, avoid parking under trees or other tall objects, as lightning tends to strike taller objects.

5. Keep your distance from other cars: If you are driving during the storm, try to keep a safe distance from other cars to reduce the risk of a collision.

6. Turn off the radio and other electronics: Turn off any electronic devices in the car, including the radio and GPS, as they can attract lightning.

7. Teach children about lightning safety: Teach your children about the dangers of lightning and what they should do to stay safe during a storm.

8. Stay calm: If you are anxious or nervous, try to stay calm and reassure your family.

9. Keep tires in good condition: Make sure that your car's tires are in good condition, as they can provide some insulation from lightning strikes.

The National Oceanic and Atmospheric Administration (NOAA) in the United States tracks lightning-related fatalities each year, including those that occur in or around vehicles. However, the specific number of people who have died from lightning while in a car is not readily available, as NOAA's statistics are not broken down by specific locations, such as vehicles.

However, according to NOAA's statistics, from 2010 to 2019, there were 128 lightning deaths in the United States, and about 64 percent of those fatalities occurred while people were engaging in outdoor leisure activities, such as camping, fishing, or hiking. Only a small fraction of lightning deaths occurred while people were in or around vehicles.

It's important to note that while lightning can and does strike vehicles, the risk of being struck while inside a car is relatively low.

The risk of being struck by lightning while flying in a commercial aircraft is also extremely low. Modern commercial airplanes are designed and equipped to safely handle lightning strikes.

When a plane is struck by lightning, the electrical charge travels along the exterior of the aircraft and is then safely discharged into the air. The passengers and crew inside the plane are protected by the Faraday cage effect, which means that the metal exterior of the plane acts as a conductor and prevents the electrical charge from entering the interior of the plane.

According to the Federal Aviation Administration (FAA), there have been no known fatalities caused by lightning strikes on commercial aircraft in the United States since the 1960s. While there have been instances of lightning strikes causing minor injuries and damage to planes, modern aviation technology has made flying in thunderstorms safer than ever before.

However, if you do find yourself in a situation where lightning strikes the plane you're traveling in:

1. Follow the instructions of the flight crew: The flight crew is trained to handle emergencies and will provide instructions on what to do during the storm. Listen to their instructions carefully and follow them.
2. Stay seated: Keep your seatbelt fastened at all times, even when the seatbelt sign is off. This will help keep you safe in case of turbulence.
3. Avoid using electronic devices: Turn off any electronic devices, including cell phones, laptops, and tablets. These devices can attract lightning and interfere with the plane's navigation system.
4. If you feel a shock, tingling, or any other unusual sensation, notify a flight attendant immediately.

5. Avoid using the restroom: Try to avoid using the restroom during the storm, as it can be dangerous to walk around the plane during turbulence.
6. Stay calm: If you are anxious or nervous, try to stay calm and reassure your family.
7. Trust the airplane: Airplanes are designed to handle lightning strikes and are equipped with lightning protection systems.

The number of people who die from lightning strikes while on a boat varies from year to year and depends on a number of factors, such as location, weather conditions, and the number of people engaged in boating activities. However, according to NOAA, in the United States, between 2006 and 2020, an average of eight people per year died from lightning strikes while engaging in boating activities.

It's important to note that being on a boat during a thunderstorm can be extremely dangerous, as boats are often made of materials that are good conductors of electricity, and bodies of water can act as a conductor and increase the risk of lightning strikes. If you're on a boat during a thunderstorm, the best thing to do is to seek shelter in a fully enclosed cabin or building on shore until the storm passes. If you can't get to shore, stay low in the boat and avoid touching metal surfaces, fishing rods, and other objects that can conduct electricity.

That said, there is always a risk of danger when in transit to safety, but as you can see, by air, land, or sea, the actual number of fatalities per year while in a vehicle or vessel is pretty small. But don't get complacent, every one of those statistical numbers was a living, breathing human being that was loved by someone, so it can and does happen; do your utmost to make sure it's not you and yours.

It's important to remember that lightning is an extremely powerful natural phenomenon that can produce electrical currents in excess of 100 million volts, so it's essential to take appropriate precautions and follow established safety guidelines. By taking the threat of lightning seriously and following these tips, you can help keep yourself, your family, and friends safe during thunderstorms.

While lightning strikes are relatively rare, there have been some unusual cases of people being struck and killed by lightning. Here are a few examples:

1. Indoor lightning strikes: Although it's rare, lightning can strike buildings and travel through electrical wiring, causing indoor lightning strikes. In 2018, a man in New Jersey was killed by an indoor lightning strike while he was working on a construction site.

2. Golf course strikes: Golf courses are a common location for lightning strikes, as they are often wide-open spaces with tall trees and flagpoles. In 1991, a group of golfers in Missouri were struck by lightning, killing one person and injuring several others.

3. Beach strikes: Being on a beach during a thunderstorm can be dangerous, as lightning can travel long distances over water. In 2014, a man in Florida was killed by a lightning strike while standing ankle-deep in the water.

4. Direct strikes: Direct lightning strikes are rare, but they can be deadly. In 2017, a man in Colorado was struck by lightning while camping in a tent and died as a result of his injuries.

5. Medical implant malfunctions: In extremely rare cases, lightning strikes have been known to cause pacemakers and other medical implants to malfunction, leading to fatalities.

STRATEGIC

For this section, I'm going to focus on lightning survival when you're in and around Fort Living Room, a.k.a. home.

When inside a home or a building during a thunderstorm,

there are several steps you can take to minimize the risk of being struck by lightning:

1. Stay inside and avoid going outside until the thunderstorm has passed.
2. Avoid using corded electronic devices, such as telephones or computers, as they can conduct electricity.
3. Stay away from windows, doors, and porches, as lightning can travel through these areas.
4. Do not take a shower or bath during a thunderstorm, as water is an excellent conductor of electricity.
5. Unplug appliances and electronics, as lightning can cause power surges that can damage or destroy them.
6. Avoid contact with metal objects, including electrical wires, pipes, and appliances.
7. If possible, stay in an interior room away from windows and doors.
8. If you live in an area prone to lightning strikes, consider installing a lightning protection system on your home or building.

Remember, the safest place to be during a thunderstorm is inside a sturdy building or a vehicle with a hardtop. Avoid open areas, tall objects, and bodies of water, as they increase the risk of being struck by lightning. If you are caught outside during a thunderstorm, seek shelter immediately in a building or vehicle, or crouch low to the ground, keeping as low as possible and making yourself as small a target as possible.

MY STORIES

When I was a boy playing out on the streets, my brother and I were caught far from home in a major summer lightning storm. We were completely stupid and playing in puddles when a bolt hit directly above our heads, blowing up the transformer and showering us in sparks. Not sure how we survived, but I know I jumped as high as I could and tried to float as long as I could before I landed back down in that puddle. We then ran into a cluster of bushes surrounded by bigger taller trees, and thankfully, the lightning did not come near us like that again, so we got lucky.

FAMILY AND FRIENDS

In these sections, I always try to provide suggestions, recommendations, and information that may be of use for those you love most. Usually, these are the things that cost a bit more than you might be willing to spend on friends, but you'll want to do your best by immediate family. While there isn't really anything that can directly protect them, here are some things to consider doing/buying for them.

1. A portable weather radio: A weather radio is a useful tool for staying informed about severe weather conditions, including thunderstorms and lightning strikes. A portable weather radio can be taken with you on the go, allowing you to stay up-to-date on changing weather conditions and potential lightning hazards.

2. A surge protector power strip: A surge protector power strip can help protect electronics and appliances from power surges caused by lightning strikes. This is a useful gift for anyone who lives in an area that experiences frequent thunderstorms.

3. A lightning detector: A lightning detector is a small device that can alert you when lightning strikes are nearby. This can be useful for outdoor enthusiasts or anyone who spends time outside during thunderstorms.

4. A hardtop metal vehicle: A hardtop metal vehicle, such as a car or truck, can offer some protection from lightning strikes. Having a reliable vehicle with a solid metal roof can help increase your safety during thunderstorms.

5. A lightning rod installation: If your loved one owns a home or building, you could consider giving them the gift of a

lightning rod installation. This can help protect their property from lightning strikes and potential damage.

While these gifts can help increase someone's safety during thunderstorms, it's important to remember that seeking appropriate shelter is still the most effective way to protect against lightning strikes. Encourage your loved ones to follow established safety guidelines during thunderstorms and help them develop a plan for staying safe during severe weather.

HOME

Since home is your castle and your family your treasure, here are some things you can do to make your home a bit more indemnified from lightning strikes. Much like one might do to safety-proof a house for a baby, there are things you can do to reduce the risk of damage or fire from lightning. In addition to the preventatives already mentioned:

1. Secure outdoor items: Make sure that outdoor items such as lawn furniture, grills, and tools are secured or brought indoors before a thunderstorm to prevent them from being damaged or becoming dangerous projectiles.
2. Trim trees: If you have tall trees near your home, have them trimmed regularly to reduce the risk of them being struck by lightning and falling onto your home.
3. Install lightning-resistant roofing: If you are planning to replace your roof, consider installing a lightning-resistant material, such as metal or tile, to reduce the risk of damage from lightning strikes.
4. Install surge protectors: Installing surge protectors on outdoor electrical outlets and on appliances or other devices in your yard can help protect them from power surges caused by lightning strikes.
5. Grounding systems: Grounding systems are designed to prevent electrical currents from lightning strikes from traveling

through your home's wiring and potentially causing damage or starting a fire. These systems can be installed by a licensed electrician.

6. Store outdoor equipment indoors: Store outdoor equipment, such as lawn mowers, bicycles, and metal tools, in a garage or other covered area during thunderstorms to reduce the risk of lightning strikes.

7. Avoid open areas: During thunderstorms, it's best to avoid open areas, such as fields or golf courses, as they are more likely to attract lightning strikes. Instead, seek shelter indoors or inside a hardtop metal vehicle.

Here are a couple of other considerations as you attempt to acquire or improvise for your safe house; some of these things you need to be aware of, as many folks may have misgivings about what works, and what doesn't.

Rubber and other insulating materials are not effective in protecting against lightning strikes. In fact, rubber soles on shoes or rubber tires on vehicles can actually increase the risk of a lightning strike by providing a pathway for the electrical current to travel through the body or the vehicle's metal frame.

Improvised materials for lightning protection are not recommended, as they may not be effective and could potentially increase the risk of injury or damage.

INTRAPERSONAL

For me personally, I've had to operate in storms with plenty of lightning, and I treat it much like I do with bullets in combat—I expect it at every turn, so that every thought and movement is with that in mind, how best to shield myself from the threat as we flow through whatever actions we're in the middle of doing against that event.

So, just using the science and facts presented here in this book and applying your own common sense, mentally and mindfully present in the moment, you can do what you need to do in the safest way possible.

MORE ON TREES

The recurring theme here is abundantly clear—if lightning reaches out and touches you directly, there is not much you can do to stop it; you can only respond and deal with it, and hopefully we've given you some good tools to use. However, if you can't find a safe haven to hunker down, and you are forced to make a movement by foot, you'll have to read the weather, the winds, speed of rain, and angle of descent (straight down is light wind, sideways rain is not).

So, for me, knowing that trees are lightning rods, and being in the open makes you one, you try to get to low ground but sometimes, especially like here in Florida where I live, near the lightning capitol of the world, Tampa, I use the trees as tools.

It is generally recommended to stay away from trees during a thunderstorm to reduce the risk of being struck by lightning. Lightning tends to strike the tallest objects in an area, and trees are often the tallest objects in a field or park.

If you are caught in a thunderstorm while outdoors, it's best to move to a low-lying area away from trees and other tall objects. If you can't get to a lower area, try to stay in an open space away from trees, and avoid being the tallest object around.

Walking from tree to tree is not recommended as it increases the likelihood of being struck by lightning, especially if the trees are taller than you. If you are in a forest, try to find a cluster of small trees or shrubs to take shelter under, but do not stand close to any single tree.

But, if I am caught out in a crazy lightning storm, I walk near enough to the trees that they will catch the lightning before me, being taller, but because trees can explode from the resin inside when struck by lightning, I stay far away enough from the tree that the explosion or falling won't cause me harm.

HAIL

I have seen hail as big as golf balls aplenty, occasionally I've seen large, egg-sized, but I've never seen the baseball- or orange-sized, though I know folks who have. Hail is a real issue and a real threat when it occurs, but the answer is always the same—seek cover.

For protections and preventions during a hailstorm, most of the same for thunderstorms apply: seek shelter, stay away from windows/glass, protect your head, stay inside your vehicle, etc.

And since hail is such an unusual thing, just for fun, I thought I'd share some real facts, history, and science on things that could also fall on your head from the sky.

There have been several instances in history when unusual objects or substances have rained down on humans. Here are a few examples:

1. Fish rain: In 2018, residents of a village in Mexico reported that small fish had fallen from the sky during a rainstorm. The phenomenon, known as fish rain, is thought to occur when waterspouts or tornadoes lift fish from bodies of water and deposit them on land.

2. Frogs and toads: There have been several historical accounts of frogs and toads falling from the sky during rainstorms. While the exact cause of this phenomenon is not fully understood, it is believed that strong winds or tornadoes may lift the amphibians into the air and deposit them on land.

3. Spiders: In 2015, residents of a town in Brazil reported that spiders had rained down on their community. The phenomenon, known as spider rain, is thought to occur when spiders release strands of silk into the air and use the wind to travel long distances.

4. Blood rain: In some parts of the world, blood rain has been reported, where rainwater appears to be stained red. This phenomenon is caused by the presence of algae or other microorganisms in the water.

5. Colored rain: In 2001, a strange-colored rain fell in the Indian state of Kerala, turning the rainwater red, yellow, green, and black. It was later determined that the cause was likely due to the presence of spores from a type of algae.

While these occurrences are relatively rare, they have captured the attention and imagination of people for centuries. And in case you weren't sure, the courses of action are the same as with hail—duck and cover!

MORE ABOUT THE SCIENCE OF WATER

Life as we know it depends on water. So, let's start with water information as that one is directly relatable by all people.

EARTH WATER FACTS:

Most folks know that our planet is about 75 percent covered in water. But most don't know that less than 3 percent of that is fresh water. Of that 2.5 percent, only 31 percent is available for use as the other 69 percent is locked up in glaciers. That means we only have .05 percent drinkable water for human consumption. Of that, only 40 percent is available to us in the form of ground water (that means underground), and that feeds the rivers and streams we depend on.

Once underground water surfaces to rivers and streams, that 1 percent provides most of our water. Of the entire Earth's mass, only .05 percent is water. The surface water which evaporates is recycled in the form of snow, rain, and other forms of precipitation, only accounting for .001 percent of our water. About 1 billion people in underdeveloped nations do not have access to adequate, clean drinking water.

That's one-eighth of the human population who live in a state of dehydration, and the resulting poor hygiene kills almost 4 million people a year by diseases caused by the unsafe drinking water. The reason they drink dirty water is because it's all they have; they then get diseases like cholera, which causes vomiting and diarrhea, leading to more dehydration and, ultimately, to their death.

So, remember, about 2.5 percent of the Earth's water is freshwater, but only a small fraction of that is readily available as drinkable water. The majority of freshwater is either frozen in glaciers or locked up in underground aquifers. Of the remaining freshwater, much of it is inaccessible or too contaminated for human consumption.

According to the United States Geological Survey (USGS), approximately 0.5 percent of the Earth's total water supply is available as freshwater that can be accessed by humans. This freshwater can be found in rivers, lakes, and groundwater sources, and only a small portion of it is easily accessible and of high enough quality for drinking.

Therefore, while the Earth has a vast amount of water, only a small percentage of it is suitable for human consumption without treatment or filtration.

Bottom line, all humans, animals, and plants survive on the less than 1 percent of water that is available to us and .2 percent of humans die each year from lack of good, clean water. Those are some startling facts.

And these are just the normal facts before we delve into potential sun-induced catastrophes that could impact it.

HUMAN WATER FACTS AND FACTORS

The average human is 65 percent water, but that range can vary plus or minus 10 percent. Fat contains less water than muscle. The water helps us with many vital health functions such as: temperature regulation, cellular function, protecting the spinal cord and other sensitive tissues, removing waste by urination, sweating, bowel movements, regulating temperature, lubricating, and cushioning joints, just to name a few.

Here's a breakdown of your organs and their relationship to water:

Body Part to Water Percentage
Brain 80–85%
Kidneys 80–85%
Heart 75–80%
Lungs 75–80%
Muscles 70–75%
Liver 70–75%
Skin 70–75%
Blood 50%
Bones 20–25%
Teeth 8–10%

The Centers for Disease Control and Prevention (CDC) recommends:
- Carrying a water bottle for easy access
- Choosing water instead of sugar-sweetened beverages
- Choosing water when eating a meal out
- Adding a wedge of lemon or lime to water to improve taste

Some tips for older adults:
- Not waiting until feeling thirst to drink fluids
- Drinking a glass of water before and after exercise
- Taking a sip of water between each bite of food at meals
- Drinking a full glass of water when taking medication

There is no set daily amount of fluid that a person should drink. The amount varies, depending on age, sex, weight, health, physical activity, and the climate where a person lives. But let's look at this a bit more since it drives your logistical planning and gives you vital data to allow you to best plan when faced with extreme situations, especially those caused by extreme weather.

Too much of a good thing? Yes. How much water is too much? Water toxicity happens when there is too much water in the body. This can dilute essential electrolytes in the blood, cause cells to swell, and put pressure on the brain.

But drinking too much water is difficult. There have been cases of water poisoning in people who drank a lot of water in a very short space of time. This may be during endurance sports because of heat stress, or when using recreational drugs that increase thirst.

There is no clear limit for drinking too much water. The kidneys can remove 20–28 liters of water per day, but they cannot excrete more than 1.0 liters per hour. Drinking more than this can be harmful.

Every day you lose water through your breath, perspiration, urine, and bowel movements. For your body to function properly, you must replenish its water supply by consuming beverages and foods that contain water.

So how much fluid does the average, healthy adult living in a temperate climate need? The US National Academies of Sciences, Engineering, and Medicine determined that an adequate daily fluid intake is:

Men: 16 cups of water a day, or roughly 4 liters
Women: 12 cups a day, or roughly 3 liters.

This of course varies widely based on age, size, weight, activity, environment, temperature, etc. The basic guideline to drink eight glasses of water a day is a great general rule that still holds true, but should vary based on the factors already mentioned.

You don't need to rely only on water to meet your fluid needs. What you eat also provides a significant portion. For example, many fruits and vegetables, such as watermelon and spinach, are almost 100 percent water by weight.

In addition, beverages such as milk, juice, and herbal teas are composed mostly of water. Even caffeinated drinks, such as coffee and soda, can contribute to your daily water intake. But go easy on sugar-sweetened drinks. Regular soda, energy or sports drinks, and other sweet drinks usually contain a lot of added sugar, which may ultimately cause dehydration by making your body get rid of more water than it needs to.

A good guide for your fluid intake is that you're probably adequately hydrated if you rarely feel thirsty or your urine is colorless or light yellow.

Drinking too much water is rarely a problem for healthy, well-nourished adults. Athletes occasionally may drink too much water in an attempt to prevent dehydration during long or intense exercise. When you drink too much water, your kidneys can't get rid of the excess water. The sodium content of your blood becomes diluted. This is called hyponatremia and it can be life-threatening.

A quick fix is a bit of salt and water to get it down. The Na from hyponatremia is the chemical symbol for salt, and hypo just means too little. Salt is one of the two main chemicals in sports rehydration drinks like Gatorade or electrolyte solutions. Therefore, you can improvise to taste your own replenishment fluids if needed. The general rule is one bottle of water, followed by one Gatorade-type drink while engaged in physical activities that produce a lot of sweating.

The effects of dehydration can vary, but the untreated end result is 100 percent death. A person may only survive a few days without water. But many factors affect how much water they need, such as their activity levels and environment. For this reason, it is not possible to tell precisely how long a person can live without water.

I'll share with you here an anecdote of facts, referencing a unique case of survival against all odds to illustrate this point: that no one can tell exactly how long a person can go without water.

There was once a man who was forced to land in the desert, flying his small plane from Los Angeles to Las Vegas. He was fine, but he had no water on board. He ate food, he worked by day when he should have rested, he did drink some urine, and had a few caffein-ated drinks and beers. Despite doing everything wrong, he survived a whopping seven days without water, for the most part, shock-ing the medical and survival experts. When asked why he thought he survived, he simply stated that he was going through an awful divorce and he refused to die and let his wife win. So, the mes-sage here is simple—you can roughly expect to only survive three days without water, but no science can measure human willpower. However, I don't recommend relying on that alone. Preparation is a far better plan.

When dehydration happens, it comes on quickly, causing extreme thirst, fatigue, and ultimately, organ failure and death. A person may go from feeling thirsty and slightly sluggish on the first day with no water, to having organ failure by the third.

I once took some young troops on a mountain hike at a brisk pace. I had a young airman walking beside me, and he instantly went from chatty to eyes rolling into the back of his head; he dropped and started doing the floppy fish. So, I juiced him with some IVs and we headed back down the mountain. The point here is that it can happen in a flash. Watch the signs.

Dehydration does not affect everyone in the same way. Each person will have a different tolerance level to dehydration and may be able to survive without water for longer or shorter periods than someone else. Some factors that determine results are age, activity levels, overall health, bodily factors such as height and weight, and sex also come into play.

What a person eats may also affect the amount of water they need to drink. For example, a person who eats water-rich foods such as fruits, juices, or vegetables, may not need to drink as much water as someone who has been eating grains, bread, and other dry foods.

The environmental conditions a person is in will also affect how much water their body uses. A person living in a very hot cli-mate will sweat, causing them to lose more water. A person in a climate-controlled environment will not sweat so they will not use as much water.

If a person who has diarrhea or is vomiting has no access to water, they will lose water much faster than someone without these issues. Remember, water impacts every single function in our bodies—regulating body temperature through sweating and breathing, aiding in digestion by forming saliva and breaking down food, moistening mucous membranes, helping to balance the pH of the body, lubricating joints and the spinal cord, helping the brain make and use certain hormones, helping transport toxins out of the cells, eliminating waste through the urine, and breath, delivering oxygen throughout the body. Without water, the body is unable to function correctly and will begin to stop working, which translates to death, plain and simple.

Here are some of the signs and symptoms to be aware of, usually starting with thirst, dehydration, and headaches. Dehydration may cause other notable changes in the body, such as: lethargy or sluggishness/lack of energy, dizziness and confusion, heatstroke and heat cramps, stiff joints that may eventually stick and not work properly, raised or otherwise unregulated body temperature, swelling in the brain, sharp changes in blood pressure, seizures and death.

When your urine starts to get dark, that's a sign you're heading in the wrong direction. When you stop sweating, you go from heat exhaustion, where you're cold, clammy and kind of blue, to heat stroke where you're hot and dry and red. This is a life-threatening emergency.

Whatever you are doing when someone hits heat stroke, stop everything, get them naked, wet and/or cold. Dip them in a river, pack them with ice, get them wet, and fan, slap cold packs in their groin, arm pits, neck, and face. If you can't do these, chances are they will die.

Again, this is not a first aid book. But I'll try to address the basics, as I am a former medical services officer, a special forces medic (like a jungle doctor) and a current national and state licensed paramedic who teaches wilderness medicine and first aid. So, if they get cold and clammy, stop and hydrate, feed, and rest. If they go hot and dry, cool them down like their life depends on it—as it does. If you can get an IV on board, do so. If they can drink, prevention is always best. If they can't drink and you can't give an IV, you can try an enema. I share a story on that in my personal section.

TACTICAL

Here we'll cover all the basics of hydration for you and your loved ones. Ensuring that your family stays hydrated during a survival situation is crucial, as water is essential for life. Here are some best methods to make sure everyone stays properly hydrated:

1. Prioritize water storage: Store an adequate amount of water for each family member, considering the general rule of one gallon per person, per day, for drinking and sanitation. Rotate and replenish your water supply every six months to maintain freshness.

2. Locate water sources: Familiarize yourself with nearby natural water sources such as rivers, streams, lakes, or ponds. In a survival situation, you may need to rely on these sources if your stored water runs out.

3. Water purification: Learn various water purification methods, such as boiling, chemical treatment (chlorine or iodine tablets), and using portable water filters or purifiers. These methods can make water from natural sources safe to drink.

4. Collect rainwater: Set up a rainwater collection system using tarps, buckets, or barrels. Be sure to filter and purify collected rainwater before drinking.

5. Educate your family: Teach your family members about the importance of hydration and how to recognize the signs of dehydration. Make sure everyone knows how to find, collect, and purify water.

6. Ration water wisely: In extreme situations, you may need to ration your water supply. Prioritize drinking water for the most vulnerable members of your family, such as young children, the elderly, and those with medical conditions.

7. Conserve water: Practice water conservation techniques like reusing graywater, minimizing water usage for cleaning, and

avoiding activities that can cause excessive sweating and dehydration.

8. Monitor hydration levels: Keep an eye on the urine color of family members, as this can be an indicator of hydration levels. Dark yellow or amber-colored urine can indicate dehydration, while pale yellow or clear urine signifies proper hydration.

9. Avoid caffeine and alcohol: These substances can have diuretic effects, which can lead to increased water loss and dehydration. Stick to water and electrolyte-rich beverages.

10. Maintain electrolyte balance: In addition to water, make sure your family has access to electrolyte-rich foods or beverages, such as sports drinks, coconut water, or homemade oral rehydration solutions. These can help replenish essential minerals lost through sweat and urine.

Remember that staying hydrated is essential for survival, and it's crucial to prioritize and plan for your family's water needs.

OPERATIONAL

For water, the biggest thing I can say is try to have enough water to survive for a full month at home for your entire family, pets, washing, clothes, cooking, etc. If you reference the eight glasses or sixteen cups a day model mentioned earlier, triple that for all the other needs per each human, and you'll have a good start point for your planning needs. Pools, wells, creeks, and cisterns are all in play as you seek to solve this logistical issue. You'll need secure routes, safe times, and good carrying capability for the distances and number of people and their condition.

STRATEGIC

For the strategic plan, this is how you plan to survive for the long haul. If you're staying put, best plan a year, or longer. If going mobile, make sure you have plenty of water resupply points along the way. One day, water may be more valuable than oil, gold, diamonds, or life.

MY STORIES

For water, I mentioned if someone got dehydrated and they couldn't drink or get an IV, then you could consider giving the person in trouble an enema. Your body can still absorb about 15 percent of the water in the bowel, and you could even use water that wouldn't be safe to drink orally.

Firstly, I teach at length in my big survival manual all about the history of urine drinking, for sexual fetishes, belief in its health and healing properties, and sometimes, for necessity.

There's a myth that you can't drink urine. This is false. You can, but it's not pleasant. There is truth to the myth *if* you are already extremely dehydrated, such as many who while lost at sea or stranded in the desert tried drinking it, but only after they had been without water a long time.

You see, the first pass or two of urine is still mainly water. But with each pass, there are more waste products concentrating in your kidneys, causing renal failure and death. So, if you have to urinate in a no-water environment, either drink it directly or save it for later, distilled or not.

And just to reinforce both the science and fact, as well as cultural resistance, NASA recycles astronaut urine (although the Russians refuse to do so).

Let's talk about many different ways to make water safe. You should try to have as many of these capabilities on you as you can. And you should do the same for everyone else you care about, for both their personal needs, if they ever got lost and separated, but also, for the whole family, if you're going to survive in a group. Things like laundry, bathing, dishes all matter a lot, so eyeball this next section and do as much as you can.

There are numerous ways to make water safe for survival. The methods vary in effectiveness, speed, and required resources. Here are some common methods to purify water:

1. Boiling: Boiling water for at least one minute (or three minutes at higher altitudes) can effectively kill most bacteria, viruses, and parasites.

2. Filtration: Using a commercial water filter or a homemade one can remove impurities and pathogens, especially if it has a pore size of 0.2 microns or smaller.

3. Chemical treatment: Adding chemical disinfectants like chlorine or iodine tablets can kill microorganisms in the water. Follow the manufacturer's instructions for proper dosage and wait time.

4. Ultraviolet (UV) light: UV light can effectively kill bacteria, viruses, and parasites. Portable UV devices are available for disinfecting water in the field.

5. Solar disinfection (SODIS): Placing clear plastic bottles filled with water in direct sunlight for six hours (or two days, if it's cloudy) can use UV radiation and heat to kill pathogens.

6. Distillation: Heating water until it becomes steam, then cooling and collecting the condensed vapor can remove impurities and pathogens. This method is effective, but requires a heat source and equipment.

7. Coagulation and flocculation: Adding chemicals like alum or natural coagulants like Moringa seeds can cause particles and impurities in the water to clump together, making them easier to remove through sedimentation or filtration.

8. Sedimentation: Allowing water to sit undisturbed for some time can cause heavier particles to settle at the bottom, while clearer water can be carefully removed from the top.

9. Cloth filtration: Pouring water through a clean cloth can help filter out larger particles and some pathogens. This method is less effective than others but can be useful in emergencies when no other options are available.

10. Natural filters: In survival situations, using natural filters like sand, charcoal, or plant materials (e.g., xylem) can help remove impurities and some pathogens from water.

Remember that using a combination of methods may be more effective in ensuring water safety. It is essential to consider the specific contaminants in the water source and the available resources to choose the most suitable method(s) for a given situation.

HOME

Having an emergency water supply is essential for survival in case of disasters or unforeseen circumstances. Here are some of the best ways to get and store water for survival at home:

1. Municipal water supply: Fill your water storage containers from the tap, as long as the water is safe to drink. Municipal water is typically treated and disinfected, making it safe for consumption.

2. Rainwater harvesting: Collect rainwater using gutters and downspouts connected to barrels or large storage tanks. Ensure the collected water is filtered and treated before consumption, as it may contain contaminants.

3. Natural water sources: If you live near a natural water source, like a river, stream, or lake, you can collect water directly from these sources. However, always filter, treat, and purify the water before using it for drinking and cooking.

4. Water wells: If you have access to a private well, ensure that the water is safe for consumption by testing it regularly and treating it as needed.

To store water for survival at home, follow these guidelines:

1. Use food-grade water storage containers: Use BPA-free, food-grade plastic or glass containers with tight-fitting lids to prevent contamination.

2. Clean containers: Thoroughly clean and sanitize the containers before filling them with water to remove any bacteria or contaminants.

3. Store in a cool, dark place: Keep your water containers away from direct sunlight and in a temperature-controlled environment to minimize bacterial growth and algae formation.

4. Rotate your supply: Periodically (every six to twelve months) rotate your water supply by consuming the stored water and refilling the containers with fresh water. This helps maintain water quality and ensures that you always have a supply of safe drinking water.

5. Label the containers: Clearly label each container with the date it was filled and any necessary treatment instructions.

6. Consider additional treatment: If you're unsure about the quality of your stored water, consider using water purification tablets, filters, or boiling it before consumption to ensure it's safe to drink.

7. Have a backup plan: In addition to storing water, have a plan for obtaining water from alternative sources and invest in water purification methods, such as portable water filters, purification tablets, or a solar still.

YARD

To ensure you have plenty of water for survival in your own yard, you can consider the following ways:

1. Install a rainwater harvesting system: This involves collecting rainwater from your roof and storing it in tanks or barrels for later use. This is a great way to have a free source of water that you can use for non-potable purposes like watering plants, cleaning, or flushing toilets.

2. Dig a well: Depending on the water table in your area, you may be able to dig a well in your yard to access groundwater. This can be a more expensive option but can provide you with a reliable source of water for drinking, cooking, and other household needs.

3. Install a graywater recycling system: This involves collecting water from sinks, showers, and washing machines, and using it for irrigation or flushing toilets. This can help you conserve water and reduce your reliance on other sources.

4. Plant drought-resistant plants: By planting drought-resistant

plants in your yard, you can reduce your water usage and conserve water for other purposes. These plants can thrive in dry conditions and require less watering than traditional landscaping.

5. Maintain water-saving practices: You can also reduce your water usage by practicing water-saving habits like fixing leaks, using low-flow fixtures, and watering your lawn and plants during the early morning or late evening hours when it is cooler and less water will be lost to evaporation.

By combining these strategies, you can ensure that you have plenty of water for survival in your own yard while also reducing your water usage and conserving this precious resource.

CAR

For your car it's good to keep a case of water. A gallon jug is good, but it doesn't allow for each person to have their own and have other backups. You'll also need something to pour the water into to avoid spreading germs. Always do the waterfall drinking method when sharing—sickness can kill in survival, prevention saves lives.

Have at least a five-point contingency plan for ways to treat water in your car, too. Fire, tablets, liquid, filter, and solar are good.

STATIC PLANNING

The basic thinking is that the amount of water an average family of four would need to survive at home without running water for one month will depend on several factors, such as climate, individual needs, and daily activities. However, as a minimum rule, the average person needs to consume at least one gallon of water per day for drinking and sanitation purposes.

Therefore, a family of four would need at least four gallons of water per day, or 120 gallons of water for the entire month, just to survive. This amount of water should be tripled for planning, if possible, to cover drinking, cooking, cleaning, hygiene, laundry, and waste disposal needs for a family of four for one month, assuming that water conservation measures are still followed.

MOBILITY PLANNING

Now, take that minimum amount of water planning number for staying home and figure out how much water you can carry in your vehicle, craft, or vessel for your number of family members and whatever other gear you're carrying, and calculate what you need and what you have. Most folks will end up with four bottles a day for seven days, or a case per family member for water amount, weight, and space as a decent planning guide.

PERSONAL

When it comes to dehydration, I have been there many times. Each time is a real challenge. But I always use common sense, move slowly, safely, and assuredly to conserve sweat and blood. Drink as much as you can when it's plentiful, and conserve but *use* when it's not.

INTRAPERSONAL

When it comes to driving myself internally, I always think of Mauro Propseri, who survived almost ten days without water in the Sahara Desert in 1994. If a human can survive that long before, they may be able to do it again, and that is what helps me push harder than the simple three-day rule. One must see beyond the established limits and see themselves there, having survived. Often enough, that will see a person through.

INTERPERSONAL

I have seen many pass out, and a few perish from dehydration. When it comes to keeping others pushing on in the face of potential death, remind them what they have to live for, find out who or what they love, and keep them driving on for that. But you should also take pauses, rest, mind the signs, and keep positive.

MISCELLANEOUS

Here are some survival tips for finding that precious life juice called water.

Look for areas that have green foliage. Green areas in a desert or plains area can indicate the presence of water. A line of trees, or some foliage low on the ground, may also hint at a creek bed or some other body of water.

You may need to go to higher ground first, at some risk of dehydration, to scout the area for signs of water. But you may get lucky and spot civilization if you're lost. Look for low-lying areas like valleys, depressions, or crevices. Water moves toward the lowest point, and rainwater may collect in these areas, too.

If you are in a mountainous area, check for water at the foot of cliffs. Try to detect signs of wildlife. Circling birds can sometimes indicate a water source nearby, or you may follow animal tracks as they often lead to water.

Check for the presence of insects. Mosquitoes are a good indicator that water is nearby, as is the presence of flies, usually staying around a hundred yards from water. Bees build hives three to five miles away from a water source.

Collect suitable snow or ice to make water. Blue hue usually means fresher water, whereas frozen salt water looks gray.

Seek water in damp sand or dirt. In a desert environment, if you spot a depression or moist patch of sand behind a sand dune, in a dried-out lake, a gully, or some similar feature, you should dig down in that area to search for water. But weigh this option carefully as digging requires a lot of energy, and a dry well could be disastrous.

Locate any water-containing plants, such as the barrel or saguaro cacti. You can also search for water in vines, plants, and roots. The general rule of thumb when checking plants and vines for water is that "clear is dear" and "white ain't right." Clear water runs from good plants, white sap from toxic ones.

Once the water is found, you'll need to make it safe. Here's a quick overview of some ways to do that. But first, let's look at the perils of drinking dirty or unpurified water. Drinking water can be poisonous or contaminated with parasites, bacteria, amoebas, and even worm cysts. It can also be contaminated by human waste. The diseases unclean water can cause are many, but there are a familiar few like dysentery, cholera, and typhoid.

If you're in survival mode while looking for water, be judicious with your sweat equity. Sweating will make you dehydrate. So be careful and use these guidelines:

1. Stay in the shade, if possible.
2. Avoid eating anything while thirsty.
3. Dampen your clothes to cool yourself in hot conditions or climates.
4. Carry materials, like a plastic sheet, to make water collection devices.
5. Be prepared to purify the water you find.

You can do this by simply bringing water to a boil over a fire, though this might not always be an option. To ensure you don't end up drinking water potentially tainted with parasites or bacteria, you should avoid stagnant water. Instead, choose water that is clear and flowing. Remember that stagnant water is an ideal breeding ground for many disease-carrying insects and parasites.

When deciding on whether or not a source of flowing water is safe to drink, know the location of the nearest settlement. The areas downstream from a city can easily be tainted by human activities. The source or origin of a water source will likely be safest for drinking.

Before you can disinfect the water, you have to find it. Depending on your location and situation, water can be abundant or virtually nonexistent. Water can come from freshwater surface sources like streams, creeks, ponds, and lakes. If you are able to distill the water, you can even use brackish or salty water as a source.

Let's not forget precipitation as an emergency water supply. Rain, snow, sleet, hail, ice, and dew can be collected for water.

Fresh rain that didn't fall through a jungle or forest canopy should be safe enough to drink as is. New snow can be melted for drinking without processing as well.

Water issuing from springs and other underground sources can also be safe in most areas. Water coming from tapped trees like maple and birch can be safe to drink and abundant in late winter. But most other water sources should be considered dirty, and must be disinfected with one of the following methods.

In order to kill the parasites, bacteria, and other pathogens in water, the most reliable thing to do is boil the water. Boiling will not evaporate all forms of chemical pollution, but it is still one of the safest methods of disinfection. One minute of a rolling boil will kill all organisms that can make humans sick from ingestion. Elevations high enough to affect boiling and cooking times will require slightly more time.

Boiling can be done over a campfire or stove in a metal, ceramic, or glass container. If no waterproof container is available, heat rocks for thirty minutes and place them into your container of water. This container could be a rock depression, a bowl burned out of wood, a folded bark container, a hide, or an animal stomach. Don't use quartz or any river rocks, as these can explode when heated.

Radiation, lead, salt, heavy metals, and many other contaminants can taint your water supply after a disaster, and trying to filter them out will only ruin your expensive water filter.

In a scenario where the only water available is dangerous water, there aren't many options. The safest solution is water distillation. Water can be heated into steam, and the steam can then be captured to create relatively clean water, despite its prior forms of contamination—including radioactive fallout.

Distillation is a great way to make water safe. But it requires some skill to operate and some special gear to make. I recommend buying one. They're fun once you have the method down, and there are many other things you can do once you know how to distill.

Water, for when you're in the field, can be produced with a solar still that collects and distills water in a hole in the ground. To build one, place a large square of a safe plastic of three to six feet over a two- to three-foot-deep hole, with a clean container centered in the bottom. (Run a drinking tube from the container so that you can drink your gathered water without taking apart the whole still.) Place dirt around the edge of the plastic at the rim of the hole to seal off the still. Place a rock in the middle of the plastic to create a 45-degree cone over the container. Dig the still in a sunny location, and in the dampest dirt or sand available. Add green vegetation and even urine to the hole to increase its water production.

A transpiration bag is a smaller and less productive version of this set-up, involving a clear plastic bag tied around live vegetation.

Water straw–type filters are a good tool for one person. Most of these filters contain an activated carbon filter element, which not only filters out larger bacteria and pathogens, but also removes odd flavors and odors from the water.

The two main types of water filters in use today are pump-action filters and drip/suction filters. The former utilizes a pump to force raw water through a filter cartridge. The latter are filter cartridges that use a gravity drip action, like an IV bag, or are placed in line on hydration bladder hoses.

When used on a hydration bladder, the user simply sucks water through the filter as needed. A good one is the Katadyn Pocket filter. It has a ceramic cartridge with silver imbedded inside. The ceramic filters out the larger pathogens, and the silver kills or disables smaller organisms like viruses. Most filters like this will pump about a quart per minute. If time isn't an issue, you might opt for a gravity-fed system.

UV light is very effective. Devices like the SteriPEN Sidewinder are battery-free, hand-cranked water purification devices that disrupt the DNA of illness-causing microbes in mere seconds. There are also a few battery-powered SteriPEN products that pack the same punch on bad bugs, and have been field-proven around the globe.

These are not 100 percent effective in water with large floating particles (which pathogens can hide behind or inside), but for clear water of questionable origin, these devices will do the trick.

Solar water disinfection (SODIS for short) is a water treatment method that uses the sun's energy for disinfection. The most common technique is to expose plastic bottles full of contaminated water to the sun for a minimum of one day.

The sun's abundant UV light kills or damages almost all biological hazards in the water. The advantages to this way of treating water are plentiful. It's easy to use, it's inexpensive or free, and it offers good (but not complete or guaranteed) bacterial and viral disinfection. Furthermore, the method uses no dangerous chemicals, and it does not require constant attention.

But there are some problems with the method. You need sunny weather or two days of overcast sky to reach the maximum effectiveness. It may be less effective against bacterial spores and cyst stages of some parasites; the water and the bottle need to be clear; only small bottles (two liters, max) can be processed.

Two of the most common and popular water disinfection tablets are Katadyn's Micropur tablets and Potable Aqua's iodine tablets. They can both work very effectively, but there are some differences.

If you're stocking up on purification tablets, you want to consider the life span of the product. The iodine tablets from Potable Aqua have a one-year shelf life. Katadyn's Micropur tablets last for two years or more. While these two products are using different chemicals, they both are better than 99 percent effective against water-borne pathogens.

Potable Aqua is faster. Water treated with these tablets is ready to drink thirty-five minutes after treatment. The Micropur tablets take a full four hours to achieve their maximum disinfecting action.

There is a potential toxicity and flavor issue with iodine. The iodine tablets are generally not a good choice for pregnant women or anyone with thyroid issues or shellfish allergies. Some folks might not be able to palate the flavor.

The Katadyn product is chlorine-based, most of which dissipates over the allotted four-hour waiting period, so that product is widely tolerated and tastes much better.

For emergencies, either bleach or iodine can be carefully used to disinfect water with good results. The amount of the chemical you use will depend on the water quality and temperature. Cold or murky water needs a little more disinfectant (four drops per quart) than warm or clear water (two drops). After adding the chemical, put the lid back on your water container and shake it for a minute.

Then turn the bottle upside down, and unscrew the cap a turn or two. Let a small amount of water out to clean the bottle threads and cap. Screw the lid back on tight, and wipe the exterior of the bottle to get the chlorine on all surfaces.

Set the bottle in a dark place, or at least in the shade, and let it sit for thirty minutes if the water is clear and at room temperature. When you open the bottle after the allotted time, it should smell like chlorine.

If not, you could add another drop or two and wait up to another thirty minutes. Best not take chances or shortcuts with water safety. The last thing anyone needs in an emergency is dysentery.

You can also use the two common forms of iodine to disinfect your water. Iodine is a more harmful substance than bleach for most people, but it is an option. To use iodine, it is critical to identify which type you are using. Tincture of iodine 2 percent is actually much stronger than 10 percent povidone-iodine solution. Use five to ten drops of tincture of iodine 2 percent in one quart of water and allow it to sit in the shade for thirty minutes.

It's always best to flush the threads and wipe down the bottle. Use five drops for clear or warm water, and up to ten for cold or cloudy water. Since 10 percent povidone-iodine solution is weaker, you'll need eight to sixteen drops per quart of water. Again, use fewer drops for good-looking water and sixteen drops for swamp water. Clean the bottle and wait.

An additional benefit to iodine is that you can use it for wound disinfections, too. Chlorine does not serve double duty like this, and you should never put bleach on any wounds. Never blend iodine and chorine for water disinfection.

Most survival books show a water filter made from charcoal-filled pants hanging from a tripod. That is better than nothing, but is not a reliable system. It will screen out larger particles, but it won't produce bacteria- and virus-free water.

But there is a cool filter that can be made from pinewood. Find a suitable piece of pinewood; look for a freshly cut piece that is at least two feet long and four to six inches in diameter. The wood should be free from knots, cracks, and insect damage.

1. Carve out the core of the wood: Use a knife or a sharp rock to carve out the core of the wood. You can start at one end, and work your way toward the other end, carving out a channel that is about one inch in diameter.

2. Create a hole in the bottom of the wood: At one end of the pinewood, drill a small hole that is slightly larger than the diameter of a pencil. This will be the outlet for the filtered water.

3. Insert a layer of charcoal: Place a layer of charcoal on top of the outlet hole. This will help remove impurities and bacteria from the water.

4. Add a layer of sand: On top of the charcoal, add a layer of sand. The sand will help remove any remaining impurities and bacteria.

5. Finish with a layer of grass or leaves: Place a layer of grass or leaves on top of the sand. This will help filter out any remaining impurities and debris.

6. Pour water into the filter: Slowly pour water into the top of the filter. The water will seep through the layers of charcoal, sand, and grass, and collect in the channel you carved out in the pinewood.

7. Collect the filtered water: Place a container at the outlet hole to collect the filtered water.

Remember to always boil or purify the filtered water before drinking it, especially if you are unsure of its quality or if it may contain harmful bacteria or viruses.

And lastly, if you have nothing at all, here are some things you can do. Drinking untreated water directly from a river, pond, or lake can be dangerous, due to the presence of harmful microorganisms, chemicals, and pollutants. However, if you do find yourself in a survival situation with no access to a water filter or bottled water, here are some ways to improve the quality and safety of the water you collect:

1. Choose your water source carefully: Look for clear, flowing water, as it tends to be cleaner than stagnant water. Avoid water sources near industrial or agricultural areas, where chemicals and contaminants are more likely to be present.

2. Collect water from the surface: Use a clean container to collect water away from the bottom and edges of the water source, as contaminants tend to settle there.

3. Let the water settle: If the water appears cloudy, or has visible particles, let it sit undisturbed for a while. The sediment will settle at the bottom, and you can then carefully pour the clearer water into another container.

4. Use a makeshift filter: If you have a piece of cloth, coffee filter, or even a sock, you can use it as a makeshift filter to remove larger particles. Pour the water through the makeshift filter into a clean container.

5. Dig a hole about one to two feet away from the body of water where the soil looks natural and fresh, versus mud or mixed soil. Make that hole one foot deep by one foot wide, and wait for it to fill as the body of water seeps through the soil, acting as a natural filter. It's not the best, but it's better than nothing.

6

Bridge Over Troubled Water

FLOODS—THE SITUATION

Floods, and the risks they bring, present themselves in many dangerous forms, the majority (but not all) due to meteorological occurrences. The bottom line here is that we are *air*-breathing mammals, and anything that compromises that simple fact of life is a clear and present danger.

It's fair to say that this particular threat is not a new kid on the block. The world in its present state has endured droughts, floods, and water-based ruination throughout both our ancient and modern history. If you doubt that, think back to an old, white-bearded guy named Noah! So, yes, there will be repetitions of what's been before, there will be events that we have not endured previously, and they may be far more prevalent than we might imagine.

Do you recall those key words mentioned within the preface? *Bigger weather and chaos.* I mention them here because increasing heat on our land surfaces and in our oceans simply converts to more water vapor in the atmosphere and, consequentially, unlooked for instances of heavy, sustained precipitation and/or torrential, even violent, downpours ushering in the chaos.

There are three main types of flooding situations that you are likely to face, namely:

1. Tidal surge/sea water inundation
2. River flooding
3. Flash flooding

Each of the above occurs due to a multitude of reasons, over and above the added peril of climate change. Some can be separated and stand alone as unique flood perils in themselves, but they all have one thing in common, and that's that each can damage or even destroy property and put lives at risk.

- Tropical storms, windstorms, ongoing/transient sea level rises and, less commonly, tsunamis and meteorite impacts can all result in seawater inundation. Perhaps the most graphic example of this in recent times was the terrible tsunami of December 2004, which created huge tidal surges across the Indian Ocean and beyond. More recently, in September 2022, Hurricane Ian was another classic example in a long list of tropical storm seawater surges, laying waste to everything in its path across Cuba and southwest Florida in particular. There will be more.

- River flooding and more general flooding occur for several reasons, heavy long or short period precipitation, hill or mountain top deluges, snowmelt, glacier melt, deforestation, poor agricultural practices, building on flood plains, and damage to dams. Across the summer of 2022, melting glaciers and overflowing glacial lakes high in the Himalayan mountains, combined with a very active monsoon to bring flood devastation to large parts of Pakistan, accounting for over 1,700 deaths. There will be more.

- Flash floods are arguably the phenomena to fear the most, given they are the most common of the three. Such floods often arrive without warning, and as the name suggests, they start and finish with rapidity. Violent thunderstorm activity is the most common cause and can result in instantaneous and debilitating flooding, even in the most unlikely places, such as arid plains or deserts. Back in August 2004, the village of Boscastle in Cornwall, England, was impacted by a hilltop deluge, which cascaded into two local rivers and ripped through the village, causing death and destruction. There are many other such examples, but there will be more, many more, often powered by record high ocean and sea temperatures, resulting in increased energy in the atmosphere, morphing into sudden deluges.

When considering the consequences such events may bring, the most obvious is a loss of life due to not being able to keep one's head above the water (i.e. drowning). Unsurprisingly, young children make up the lion's share of the deaths in this case. However, other equally perilous dangers may also present themselves. Hypothermia is a classic consequence of being in the water for too long and losing vital body heat; while a variety of objects carried within fast-flowing water could be lethal if they strike with enough force. Other nasties that may present themselves include poisonous or hungry or distressed wildlife, such as the alligator that got transported to your yard! Then there's the very real risk of electrocution, given that water is a perfect conduit of the electricity you may not have had the time to turn off at the mains. Finally, though the peril list will undoubtedly be longer than I have stated here, in the aftermath of flood disasters, a lack of clean drinking water and the emergence of diseases such as cholera and typhoid may present themselves and be every bit as life endangering as the flood itself.

What's critical in all of this is where you are located or where you might find yourself, with concern for the topography and type of hazards you are most likely to face. More often than not, flood-related occurrences are either localized or regionalized, with height above sea level being one of several critical factors. And sea levels are on the rise. Greenland, Antarctica, the Arctic, and the world's massive permafrost zones are gradually giving up their snow and ice, all being converted to liquid water for transportation around the globe, which means many populations will be seeing their flood risk rise immeasurably over the decades to come. By the way, it's not just that tiny group of low-lying islands many thousands of miles away you may have heard about on the news—it includes major cities. Here, care of earth.org, is the current top ten at-threat cities that may be sinking below the waves within the next one hundred years . . . and perhaps much sooner.

1. Bangkok, Thailand
2. Amsterdam, Netherlands
3. Ho Chi Mihn City, Vietnam
4. Cardiff, UK
5. New Orleans, US

6. Manila, Philippines
7. London, UK
8. Shenzhen, China
9. Hamburg, Germany
10. Dubai, UAE

So, there we go, troubled water by the bucketful. We need a bridge or perhaps more likely a life jacket!

FLOODS—THE SURVIVAL

For eons, floods have shaped our landscapes, molded civilizations, and impacted human history. From the biblical tale of Noah to the inundation of great cities, they've carved their mark on both the physical world and the annals of humanity. Respect and understanding of these powerful forces are paramount.

THE SCIENCE

Floods occur when water overwhelms a land area that is usually dry, resulting from a variety of factors. These include heavy rainfall, rapidly melting of snow, tsunami waves, or the collapse of natural water containment systems like dams. Factors such as deforestation, urban development, and climate change can exacerbate the impact of floods by altering natural drainage systems and increasing the volume and speed of runoff.

THE DIFFERENT TYPES OF FLOODS

Riverine Floods: Caused by overflows from rivers, these are especially common in areas with large river systems and frequent heavy rains. Regions like South Asia and the US Midwest experience them regularly.

Coastal Floods: Resulting from storm surges, these are often accompanied by heavy rainfall and can affect coastal regions. Places like the US East Coast and Southeast Asia are at high risk.

Flash Floods: These are sudden, intense floods caused by heavy rain in a short period, especially in areas with steep terrain. They can occur anywhere, but are common in regions like the American Southwest.

Pluvial Floods: Occur when heavy rainfall leads to surface-water flooding in flat areas where the ground becomes saturated.

Keys include: meteorological warnings or forecasts of heavy rains; rapidly rising water levels in local waterways; changes in water color or increased debris in rivers and streams.

To prepare for anticipation of flooding, which is often part of extreme storm events and factored into applicable warnings, it is paramount to stay continuously informed as these conditions can change rapidly. The same preventions and safety measures that are recommended for severe storms apply here, such as storing essential food, water, medicines, and other supplies; protecting important documents and making sure you have adequate insurance (note that many insurance policies do *not* include flood insurance, so if you live in a flood-prone region, it would be wise to have a policy for flood damage).

Also, as with hurricanes and other such events, have an evacuation plan and routes and practice them in advance.

DIFFERENCES AND CAUSES

Tsunamis, tidal waves, and other types of flooding are all related to the movement of water, but they differ in their causes, mechanisms, and impacts. Let's explore the differences between these phenomena:

1. TSUNAMIS
Difference: A series of powerful waves with long wavelengths. They can be very high, moving inland rapidly.

Mechanism: When the seafloor is suddenly lifted or dropped, it displaces a massive amount of water, creating a series of waves that radiate outward from the disturbance's epicenter.

Causes: Underwater earthquakes, volcanic eruptions, or landslides.

Indicators: Sudden receding of ocean water, a loud roar from the ocean, noticeable rapid rise or fall in coastal waters.

2. TIDAL WAVES
Difference: Often confused with tsunamis, but tidal waves are related to the gravitational interactions of the Sun, Moon, and Earth.

Mechanism: Tidal waves are primarily caused by gravitational interactions and the Earth's rotation, leading to the rhythmic movement of water in coastal areas.

Causes: Astronomical influences and shifting atmospheric conditions.

Indicators: Predictable based on lunar cycles, not typically destructive or noticeable unless combined with other meteorological factors.

3. STORM SURGES/FLOODING
Difference: A temporary rise in sea level due to strong winds from a storm, especially hurricanes, and low atmospheric pressure.

Causes: Severe weather systems like cyclones, hurricanes, or typhoons.

Mechanism: The causes of flooding are diverse, but they generally involve the accumulation of water beyond the capacity of natural or man-made drainage systems.

Indicators: Meteorological warnings, dark, rapidly lowering, and thick cloud formations, rising water levels in coastal areas.

4. LOCALIZED FLOODING

Urban Flooding: Result of inadequate drainage systems in urban areas. Occurs during heavy rainfall, prevalent in densely populated regions.

Dam and Levee Failures: Result from structural issues or excessive water pressure. Can occur anywhere with dams or levees.

SURVIVAL TACTICS

HOME

- Move to higher ground or the highest level of your home.
- Stay away from windows to avoid breaking glass.
- Have an emergency kit ready.

VEHICLE OR BOAT

- Don't try to outrun the tsunami, especially if you're close to the shore.
- Leave the vehicle and move to higher ground as fast as you can.
- Head out to deeper water if there's time.
- If you're in shallow water or harbor, abandon the boat and seek higher ground.

ON FOOT

- Drop everything and move to higher ground immediately.
- Avoid downed power lines and stay away from buildings and trees.

EARLY WARNING TECHNOLOGIES

- Seismic Sensors: Detect underground seismic activity that may lead to tsunamis.
- Deep-ocean Assessment and Reporting of Tsunamis (DART): Measures changes in sea level in the open ocean.
- Tide gauges: Monitor sea levels to identify any abnormal rises or drops.

- Satellites: Can detect and measure tsunamis in deep ocean areas.
- Early warning systems (EWS): Combine data from various sources to predict and alert areas at risk.

GADGETS FOR TSUNAMI SURVIVAL

- Emergency radios: Receive updates and warnings even if power is out.
- Life jackets: Increase chances of floating in floodwaters.
- Personal locator beacon (PLB): Sends out a distress signal with your location to rescue services.
- Waterproof flashlight: Essential for visibility during and after the event.
- Water purification tablets/devices: Access to clean drinking water if supplies are contaminated.

EMERGING TECHNOLOGIES FOR TSUNAMI SURVIVAL

- AI and machine learning: Predict tsunamis more accurately by analyzing vast amounts of data quickly.
- Improved building materials: Structures that can absorb tsunami energy or float above floodwaters.
- Drones: For real-time monitoring, assessment, and even delivery of essential supplies during post-tsunami scenarios.
- Virtual reality (VR): For better preparedness through simulation-based training on tsunami scenarios.
- Communication technologies: Improved networks to ensure uninterrupted communication during and after the event.

In any scenario involving tsunamis or large-scale flooding, preparedness and education are vital. Being informed, having a plan, and understanding the environment around you can significantly increase your chances of survival.

STRATEGIES
The same preventatives for other extreme weather events apply here, including preparedness involvement of the community,

maintaining a stockpile of appropriate supplies, fortifying and weatherproofing your home and property, keeping your vehicles fueled and properly maintained, etc.

Know the local high ground and the quickest routes to them. Always move perpendicular to floodwaters. When traveling in a vehicle, avoid water-covered roads; twelve inches of water can sweep away most cars. Know alternate routes.

HISTORICAL PERSPECTIVES

Ancient floods:
- The Great Flood (Mythological accounts worldwide)
- The Huang He (Yellow River) floods, China
- St. Marcellus's flood, Netherlands
- St. Lucia's flood, Netherlands
- The 1099 flood of the Ganges and the Brahmaputra, Bangladesh

Modern disasters:
- 1931 China floods
- 2004 Indian Ocean tsunami
- 2010 Pakistan floods
- 1953 North Sea flood
- 1970 Bhola cyclone

Annually, floods result in the deaths of approximately 7,000 people worldwide.

While floods are powerful and often devastating, human resilience and preparedness are even more formidable. Every year, countless individuals brave these waters and emerge stronger, often with the support of their communities. With knowledge and preparation, you, too, can navigate the stormy tides and safeguard not just yourself, but those around you. Remember, it's not the waters we face, but how we face the waters that define us.

SECTION IV
WIND AND STORMS

I know I'm alive when I feel the wind on my face. I know I'm dead when I feel the wind in my head.

Riders on the Storm

HURRICANES AND TYPHOONS— THE SITUATION

And now we arrive at a big place and an equally big chapter, where we explore the monsters of the globe and the potential devastation and distress each may bring. Hurricanes, typhoons, cyclones, and storms generally possess various weapons of destruction, not only on a local or regional level, but also on a far wider reach that often takes in large swaths of the Earth's surface.

I say various weapons, because the deadly arsenal of these storms from the mouth of hell contains everything from stupendous rain, catastrophic flooding, seawater inundation and deadly tornadoes, to debilitating sandstorms, huge thunderstorms, violent winds and, more than anything else, truly life-threatening, mountainous seas. There may well be other exotic and less exotic threats to your life, limbs, and property tucked away somewhere within the cores of these giants, but in this chapter, we are going to focus on the final two in that list, namely destructive winds and the gigantic seas that can be stirred up when the demons come out to play.

First of all, a little more physics for you, and I promise not to go too deep (he says biting his bottom lip!). Before we get to our topic, water . . . let's discuss wind.

So, what exactly is wind, and why does it happen at all? Well, most things in life begin and end with the Sun having a hand in matters, and the materialization of wind is no different—it's the mainstay. The Earth's surface is not warmed evenly from one place to the next, nor from day to night, whether we are talking about the local park and small microclimates within, or the huge insolation difference between say the Arctic and the Equator; the laws of physics apply equally. Both of these examples, and everything in between, are a cause for air pressure differentials, with the areas of greatest heating tending to develop into lower air pressure zones, while higher air pressure zones tend to frequent cooler or colder zones. The advent of a sea breeze is a perfect example of that mechanism in play at a local level. However, things are never that simple, and the expected trends are often misaligned or broken. Otherwise our weather would be pretty well predictable—and I might not have a job!

Instead, we have a constant spinning jenny on our hands, namely the Earth's rotation. We also have another force playing a part in compressing air molecules from above, namely gravity. It's a complex and highly energetic amalgamation, both on the surface, and much higher up toward the ceiling of our livable atmosphere known as the tropopause, or the lid on our weather.

The bottom line here is that akin to most planets in our solar system, we have a truly dynamic global system, one that never sits still and one in which higher air pressure areas are forever attempting to move toward lower air areas, striving to achieve the natural balance that nature and physics demand. But, as these air masses move, they are being constantly deflected by the rotation of the Earth, changing and metamorphizing as they careen across oceans and landmasses, warming and cooling as they do so, and yes, occasionally developing into cyclonic monsters. On the other hand, it all might transpire in a far more modest way within the limits of your local park, or even around that piece of litter you might see dancing about in the street; it's the same physics at work. That is what we call wind. By the way, similar forces are at play within your digestive system, for example, but we won't go into that!

So, now we know what that force you feel on your face or in your hair is all about, allow me to provide some meat on the bones concerning those cyclonic monsters I mentioned, the ones that can bring anxiety, death, and destruction in equal measure.

Hurricanes, typhoons, and cyclones are, in fact, one and the same thing—massive, rotating tropical storms, their label dictated by which ocean they happen to be spawned in. For hurricanes, it's in and around the Caribbean, within the warm waters around the Cape Verdean islands of West Africa, the east coast of Central America, the Gulf Coast of the southern United States, and the eastern North Pacific. For typhoons, the spawning ground is in the western Pacific Ocean, while cyclones begin their lives in the northern Indian Ocean and also within the southern hemisphere and rotating in an opposite clockwise direction within the southern Indian, southern Pacific, and more rarely, the south Atlantic oceans. Those northern hemisphere storms are normally on the march between the months of June and November, while their southern counterparts are generally present between November and April, though almost always less numerous in number.

The common thread for all of these storms is that their birth is dependent on ocean temperatures being above 26.5 degrees Celsius, heated by the intense overhead tropical sun, allowing hot and humid air to rise and form into depressions. Given enough of the necessary fuel, the depressions will rapidly deepen and begin to spin. From that point, there will be a high degree of unpredictability and volatility as to its development and direction of travel—and let's be clear, no tropical cyclone has exactly the same characteristics as any of those that went before; they are all unique in their own majestic and frightening way.

The larger tropical depressions of every variety are awarded a name set by the World Meteorological Organization, once sustained wind speeds reach 39 mph, though that accolade may not necessarily be appropriate given what follows if the tropical storms continue to deepen to a point where the sustained wind speeds reach 74 mph and turn into hurricanes, typhoons, or cyclones. From that point forward, the storms assume their erratic track toward their various destinations, occasionally (but far from always) wreaking havoc, death, and destruction to villages, towns, cities, and whole countries alike.

There is, however, some good news about this particular phenomenon in that, due to satellite and radar technology, meteorologists like myself are able to track these beasts from their birth to their demise, more often than not allowing for ample warning and evacuation. Also, at the point of this writing, there is not yet any solid scientific proof that climate change has caused these storms to be either more numerous, more powerful, or able to extend their lives regularly out of their normal season—but watch the space, as I feel it will soon be filled!

This is where the good news ends, because if such storms pay you a visit, then you certainly will be at the mercy of the elements, with the capacity for destruction measured case as follows, care of the National Hurricane Center and the Central Pacific Hurricane Center.

The Saffir-Simpson Hurricane Wind Scale is a 1 to 5 rating, based only on a hurricane's maximum sustained wind speed. This scale does not take into account other potentially deadly hazards such as storm surge, rainfall flooding, and tornadoes.

The Saffir-Simpson Hurricane Wind Scale estimates potential property damage. While all hurricanes produce life-threatening winds, hurricanes rated Category 3 and higher are known as major hurricanes. Major hurricanes can cause devastating to catastrophic wind damage and significant loss of life, simply due to the strength of their winds. Such hazards require people to take protective action, including evacuating from areas vulnerable to storm surge.

In the western North Pacific, the term super typhoon is used for tropical cyclones with sustained winds exceeding 150 mph.

Category	Sustained Winds	Types of Damage Due to Hurricane Winds
1	74–95 mph 64–82 kt 119–153 km/h	**Very dangerous winds will produce some damage:** Well-constructed frame homes could have damage to roof, shingles, vinyl siding, and gutters. Large branches of trees will snap and shallowly rooted trees may be toppled. Extensive damage to power lines and poles likely will result in power outages that could last a few to several days.

(Continued on next page)

Category	Sustained Winds	Types of Damage Due to Hurricane Winds
2	96–110 mph 83–95 kt 154–177 km/h	**Extremely dangerous winds will cause extensive damage:** Well-constructed frame homes could sustain major roof and siding damage. Many shallowly rooted trees will be snapped or uprooted and block numerous roads. Near-total power loss is expected with outages that could last from several days to weeks.
3 (major)	111–129 mph 96–112 kt 178–208 km/h	**Devastating damage will occur:** Well-built framed homes may incur major damage or removal of roof decking and gable ends. Many trees will be snapped or uprooted, blocking numerous roads. Electricity and water will be unavailable for several days to weeks after the storm passes.
4 (major)	130–156 mph 113–136 kt 209–251 km/h	**Catastrophic damage will occur:** Well-built framed homes can sustain severe damage with loss of most of the roof structure and/or some exterior walls. Most trees will be snapped or uprooted and power poles downed. Fallen trees and power poles will isolate residential areas. Power outages will last weeks to possibly months. Most of the area will be uninhabitable for weeks or months.
5 (major)	157 mph or higher 137 kt or higher 252 km/h or higher	**Catastrophic damage will occur:** A high percentage of framed homes will be destroyed, with total roof failure and wall collapse. Fallen trees and power poles will isolate residential areas. Power outages will last for weeks to possibly months. Most of the area will be uninhabitable for weeks or months.

So, what countries are statistically thought to be most at risk from tropical storms? There are numerous candidates, and many endgame destinations occupy locations thousands of miles away from where storms were first conceived. For the record, typhoons are more common than cyclones which are, in turn, more common than hurricanes. However, if there is a top ten hit list (and the hits come in various forms) then this is it (subject to future changes of course):

1. Southern China
2. The Philippines
3. Japan
4. The USA (North Carolina being the most prone US state)
5. Mexico
6. Vietnam
7. Australia
8. Madagascar
9. India (Puri being the world's most prone city)
10. Laos

Beyond Puri, which is a coastal city some two hundred miles south-west of Kolkata in India, there are only a few places that on *average* can expect more than one tropical storm per season, but there are some close calls in the US. These include, for example, Cape Hatteras and Morehead City in North Carolina, which average a tropical storm every 1.3 to 1.5 years, respectively. As for my home continent of Europe, tropical storms in the true sense of the word hardly ever appear on the menu. However, there have been two modern-day hits, namely Hurricane Vince in 2006, which crashed into southwest Spain as a tropical depression, and subtropical Storm Alpha in 2020, which swept into northwest Portugal at full force. All other storms to impact Europe, bar a couple of disputed examples spanning more than a century, were neither tropical nor subtropical in nature, but that does not negate the fact that severe and dangerous storms have ravaged almost every part of Europe and every other continent and ocean of our green and blue globe since the beginning of all things.

Talking about beginnings, my company British Weather Services was formed immediately in the aftermath of the UK's greatest modern-day storm, in October 1987. This particular storm was a sudden cyclonic event that the government's meteorological office failed to forecast. It went on to kill eighteen people and losses totalled £1.4 billion ($2.5 billion US). Thankfully, it was an overnight occurrence, otherwise The Great Storm, as it became known in the aftermath, would have been far more costly in lives lost and the damage wreaked.

It's fair to say that you may well have experienced a time of stormy winds of one velocity or another, perhaps a time when the wind speeds were strong enough to blow you off your feet, or to pick up and carry objects aloft at tremendous speeds, or even to rip buildings apart. Most of us have been there, and just as the Earth continues to spin, so will these windstorms continue to evolve and threaten.

During my time with the British Royal Navy, I experienced and endured hurricane-force winds and windswept, tumultuous seas on many occasions. I vividly recall a day of mountainous swells created by a huge cyclone off Uruguay, a time when onboard the old colonial ship SS *Uganda* (look this one up, if you care to), in the south Atlantic on the way to the Falkland Islands. The ship pitched and lurched in and out of unimaginable deep troughs and high peaks, enough to send many of my fellow sailors to their bunks in distress. I also remember my time in the Southern Ocean (the world's windiest ocean) onboard RFA's *Fort Grange* and *Fort Austin* (these old grey ships are also worth investigating), between the Falklands and the tiny, beautiful but raw island of South Georgia, located within the Antarctic convergence zone. By the way, Antarctica is the windiest continent on Earth. It was there, while iceberg spotting on the bridges of those support ships, that my shipmates and I constantly faced what every round-the-world yachtsman has to overcome in terms of turbulent, wind-driven, foamed-filled seas, as terrible and rough as they come. And during my time there, while anchored off Cumberland Bay in South Georgia, on a peaceful moonlit evening, I was sorely tested, and so was the ship, by the sudden advent of an 80 mph katabatic wind racing down off the huge ice mass of the Fortuna Glacier which, within minutes, had the ship helplessly careering around like a maggot on a fly-fishing line. Thankfully, once the air pressure had equalized, the terrible winds died down almost as fast as they had arrived, all of which takes us back to that physics lesson near the beginning of this chapter. My own experiences were just a tiny taste of what every seafarer through the ages would testify to, very likely with a large dose of apprehension, fear, and of course, respect.

Finally, for now, I'll mention here something I have personally found to be very useful, not only in my various meteorological tasks, but also for whenever the wind blows in and out of my job—namely the Beaufort wind force scale, a scale first constructed by British Admiral Sir Francis Beaufort back in 1806 for use at sea. It's since been adopted for use on land and there are various versions out there to be found. Discover one that you like the look of and have regard for what each wind speed might mean in the environment you find yourself in. For example, (and remembering that it's the gusts that lead to most concern/damage opposed to the sustained winds), a Beaufort Force 5 fresh breeze of 19–24 mph will see tree branches swaying and crested waves appear on inland waters; while a Gale Force 8 wind will certainly impede your walking progress, could well see small branches break from trees and expect to observe trash cans and the like careering down the street.

HURRICANES AND TYPHOONS—
THE SURVIVAL

Jim does a great job of covering the science for this chapter, so I won't delve too deeply there. I'll start with a bit of interesting relevant facts, some extreme history, and a bit of wind trivia. My purpose is simple, to get your attention, and hopefully, a new-found respect for the power of wind.

FACTS

Extremely high winds have blown down many things to include:

1. Trees: High winds can cause large trees to fall, even uprooting them completely.
2. Buildings: In areas where structures are not built to withstand high winds, they can be blown down or severely damaged.
3. Cranes: High winds can cause construction cranes to topple over, potentially causing serious damage and injuries.
4. Airplanes: Strong crosswinds during takeoff and landing can cause airplanes to lose control and crash.
5. Cars: High winds can also blow cars off the road or cause accidents.

6. Power lines and poles: Strong winds can cause power lines and poles to collapse, leading to power outages and potentially dangerous situations.
7. Ships: High winds can cause ships to capsize or run aground.
8. Signage: Large and heavy signs, especially those that are not properly secured, can be blown down by high winds.
9. Tents and temporary structures: Tents, canopies, and other temporary structures can be destroyed by high winds, potentially causing injuries to those inside.
10. Fences: Even sturdy fences can be knocked down by extremely high winds, especially if they are not properly anchored.

TRIVIA

While extreme winds can blow down many common items like trees, buildings, and power lines, they can also cause some strange things to happen. Here are some examples:

1. In 2018, a powerful storm in Turkey caused a parked Boeing 737 airplane to be blown off its chocks and into a nearby building.
2. An historic lighthouse on the Outer Banks of North Carolina was toppled in 2018 by Hurricane Michael.
3. Strong winds in California's Death Valley National Park blew a 2,000-pound boulder off its perch and sent it rolling down a hill in 2015.
4. A Ferris wheel in Myanmar was blown down by strong winds during a festival in 2017, injuring several people.
5. In 2020, a trampoline was blown into a tree by high winds in the UK, creating a bizarre and dangerous sight.
6. High winds caused a hot air balloon to crash into a power line in Massachusetts in 2016, leaving its passengers dangling in the air.
7. In 2019, strong winds in Colorado blew a portable toilet across a park, causing some amusement and disbelief among onlookers.

8. A 40-foot inflatable rubber duck blew off its moorings in Hong Kong's Victoria Harbor during a 2013 storm, causing it to float away.

HISTORY

Extreme wind events can cause a range of disasters, from property damage to loss of life. Here are some of the worst disasters caused by extreme wind:

1. Hurricane Katrina: In 2005, Hurricane Katrina caused catastrophic damage along the Gulf Coast of the United States, particularly in New Orleans. The storm surge and high winds led to over 1,800 deaths and $125 billion in damage.

2. Cyclone Nargis: In 2008, Cyclone Nargis struck Myanmar, causing catastrophic damage and loss of life. The storm, which had sustained winds of 120 mph, resulted in over 138,000 deaths and $12 billion in damage.

3. Super Typhoon Haiyan: In 2013, Super Typhoon Haiyan struck the Philippines, causing widespread destruction and loss of life. The storm, which had sustained winds of 195 mph, resulted in over 6,300 deaths and $2.2 billion in damage.

4. The Great Galveston hurricane: In 1900, the Great Galveston hurricane struck the city of Galveston, Texas, causing widespread destruction and loss of life. The storm, which had sustained winds of 145 mph, resulted in over 6,000 deaths and $30 million in damage (equivalent to over $1 billion in today's dollars).

5. Cyclone Idai: In 2019, Cyclone Idai struck Mozambique, causing catastrophic damage and loss of life. The storm, which had sustained winds of 115 mph, resulted in over 1,000 deaths and $2 billion in damage.

6. The 1974 Super Outbreak: In 1974, a massive outbreak of tornadoes struck the central and southeastern United States, resulting in 148 confirmed tornadoes and over 300 deaths.

7. The Edmonton tornado: In 1987, a powerful tornado struck the city of Edmonton, Alberta, Canada, causing widespread destruction and loss of life. The tornado, which had winds of up to 300 km/h (186 mph), resulted in 27 deaths and $500 million in damage.

TACTICAL

Now that I have your attention, let me talk about some things you can do at the tactical level which, in this case will be as a human, without a vehicle or building for protection.

If you find yourself caught out in extreme wind with no shelter, it's important to take immediate action to protect yourself. Here are some tips:

1. Seek shelter: If possible, try to find a nearby building or other structure to take shelter in. Look for a sturdy building with a basement or interior room on the lowest floor.

2. Stay low: If you cannot find shelter, try to get as low as possible. Get down on the ground and crawl to a low spot or depression where the wind may be less intense.

3. Protect your head: Cover your head with your arms or with a sturdy object, such as a backpack or a piece of debris. This will help protect you from flying debris.

4. Stay away from trees and power lines: Avoid standing under trees or near power lines, as these can be dangerous during high winds.

5. Cover your mouth and nose: Use a scarf or other piece of cloth to cover your mouth and nose, as blowing dust and debris can be harmful to your respiratory system.

6. Wait it out: If you can't find shelter, or a safe place to wait out the storm, huddle against a sturdy object, such as a building or a large rock, and wait until the wind dies down.

Remember that extreme wind can be very dangerous, so it's important to take any available measures to protect yourself. If you have access to a weather radio or other source of information, listen for updates and follow any instructions from local authorities.

OPERATIONAL

If you are caught out in extreme wind in a car, in the middle of nowhere, basically the same preventions and safety measures apply as those discussed for storm events, such as pulling over and stopping, keeping your seatbelt on, keeping windows rolled up, staying away from power lines, etc.

Remember that extreme wind can be very dangerous, so it's important to take any available measures to protect yourself. If you have access to a phone or other communication device, call for help if you need it.

If you are caught out in extreme wind in a small boat:

1. Lower the sails: If you are sailing, lower the sails as quickly and safely as possible. This will help reduce the boat's wind resistance and prevent it from capsizing.

2. Use the engine: If your boat is equipped with an engine, use it to maneuver the boat into a safe position. Try to keep the bow of the boat facing into the wind and waves.

3. Wear life jackets: Make sure everyone on board is wearing a life jacket or personal flotation device.

4. Stay low in the boat: If possible, stay low in the boat to reduce your exposure to the wind and waves. Get down into the cabin if you have one.

5. Use an anchor: If you can find a safe spot, drop anchor to keep the boat from drifting.

6. Call for help: If you have access to a communication device, such as a VHF radio or satellite phone, call for help if you need it.

7. Stay with the boat: If you need to abandon the boat, do so only as a last resort. In most cases, it's safer to stay with the boat, as it will be easier for rescuers to spot.

If you are caught out in extreme wind in a small plane:

1. Try to fly to a safe location: If possible, try to fly to a safe location with less extreme wind conditions. Look for a nearby airport or landing strip that you can safely reach.
2. Use your instruments: If you are flying in low visibility, use your instruments to navigate and maintain altitude. Stay on course and avoid overcorrecting.
3. Communicate with air traffic control: If you have access to a radio, communicate with air traffic control for assistance and guidance.
4. Reduce speed: Reduce your airspeed to help maintain control of the plane in high winds.
5. Try to find a clear area: Look for a clear area with no obstructions, such as trees or power lines, to land the plane.
6. Approach with caution: If you are landing in high winds, approach with caution and make sure your landing gear is down. Use a shallow approach angle to reduce the chance of stalling.
7. Brace for impact: If you are unable to safely land the plane, brace for impact and try to protect yourself as much as possible.

For protection in the home against wind events, the same preventions and safety measures apply as those discussed for storm events—such as good yard maintenance, gutter and roof upkeep, securing of outdoor objects and furniture. In addition:

1. Reinforce doors and windows: Make sure that all doors and windows are properly secured and reinforced. Install storm shutters or plywood panels to cover windows and doors during high winds.
2. Install hurricane straps: Install hurricane straps or clips to secure the roof to the walls of your home to reduce the risk of roof damage during high winds.

3. Reinforce your garage door: Make sure your garage door is reinforced and can withstand high winds. Install additional bracing if necessary.

4. Check your insurance: Review your insurance policy to make sure you are adequately covered for damage caused by high winds.

Remember that extreme wind events can be very dangerous, so it's important to take any available measures to protect your home and your family. Listen for updates and follow any instructions from local authorities during severe weather events.

If you live in an area that is prone to hurricanes and tornadoes, the same permanent measures advised for storms apply here, such as the installation of impact-resistant windows and doors, metal roofing, a storm shelter or safe room, etc.

MY STORIES

On September 22, 1998, Hurricane Georges hit Haiti. I was there working for the State Department as the country manager of the only aviation in country. We had two Russian Mi-8, cargo/passenger helicopters with Russian crews. We had an Argentine QRF (quick reaction force), and we flew the Haitian leadership, the United Nations, and, of course, the US Embassy and military group, so I spoke English with the embassy, French with the Haitians and UN, Spanish to the QRF, and Russian with my Russian Pilots . . . good times.

We knew the hurricane was coming, so we had to pack up everything and fly to Aruba, outside the hurricane belt. We knew there might not be anything left, so we loaded our radio gear, our clinic, our night vision and weapons, all our helo gear, and the extra fuel bladders. We were going to try to fly there on the gas we had, as there was nowhere to stop, many hours over nothing but ocean. We made it. The storm hit. We went back.

We recovered eighty-three bodies. Over six hundred died. More structures survived than we expected. We spent the next few months helping with the recovery from the damage. The Haitian people were resilient and spirited, despite their calamity. I could not help but admire them. I now have a young man I'm honored to call a friend who is Haitian. He is a marine raider, and one of the finest men I've ever known. He is a descendent of Papa Doc. Those who know, know. But his heroism on the battlefield saving so many marine lives under fire, and enduring the racism he has, makes me think he has well redeemed his family's honor.

We lost thirteen during that period, when a chopper on a lifesaving emergency med evac in a bad rainstorm crashed into a mountainside. I was awarded Small Contractor of the Year.

PERSONAL

In my personal life, I have lived through more than a dozen hurricanes, including Irma in Florida, September 11, 2017. One of the red overturned vehicles on the front page of ABC News was a vehicle I helped recover.

Most recently, on September 28, 2022, I was there when Category 4 Hurricane Ian hit, and it was a fearsome thing. My wife, Ruth, survived Hurricane Isaac in New Orleans, Louisiana, in August 2012, and she was my hero, because I was away and she was there alone, with our very sick young son of six.

She could not escape, as she had only recently moved to America, and when she went to buy a car here, she insisted on a sedan—against my recommendation for an SUV. The floods came so suddenly, there was no way her car could get out of the driveway. She managed all like a champ, without power or water for days in sweltering heat with a highly febrile lad with a 104 temp. She saved his life with her care.

FAMILY, FRIENDS, AND NEIGHBORS

Outside of the staying in the house, car, or outdoor alone measures, the same preventions and safety measures for storms also apply here, such as having shelter and communication plans, having emergency kits, food and water, supplies, etc. In addition, make sure your insurance coverage is up to date with adequate coverage for wind damage; also, have an evacuation plan, should that become necessary.

MOBILITY PLANNING

While static planning for wind events is much the same as for storms, there are a few things which bear mentioning or reiterating for mobility. Here, I can honestly say that I hope no one with their God-given common sense would ever venture out into a hurricane by car or foot. But sometimes, you might have an isolated child that you simply must get to, and that would be a worthy cause to risk your life for—because you will be 100 percent risking your life. That said, in my mind I think of ten miles as a good distance in a storm or jungle, twenty-five miles in a day, for flat lands. So, if you have to make that trek by foot, here are a few tips.

1. Wear appropriate clothing: Wear suitable clothing, such as waterproof boots, rain gear, and warm clothing, to protect against wind, rain, and cold temperatures.
2. Pack a backpack: Pack a backpack with essentials like food, water, first aid supplies, a flashlight, and a battery-powered radio. Make sure to include enough supplies for the entire journey.
3. Stay informed: Stay informed about weather conditions and warnings by listening to weather reports on the radio or television, or by using a weather app on your phone.
4. Plan your route: Plan your route in advance and avoid low-lying areas, as they are more prone to flooding.
5. Walk against the wind: Walk against the wind to reduce the risk of being blown off balance.
6. Stay low: If the wind becomes too strong, get down on the ground and crawl to a low spot or depression where the wind may be less intense.
7. Use caution around debris: Use caution around debris and avoid stepping on sharp objects or broken glass.

8. Rest frequently: Take frequent breaks and rest when needed to avoid exhaustion.
9. Signal for help: If you need assistance, use a whistle or other signaling device to attract attention.

Remember that walking anywhere in the middle of a hurricane is a last resort and should only be done if necessary. If possible, seek shelter and wait until the storm has passed before attempting to travel. Stay informed and take all available measures to protect yourself from the dangers of the storm.

MISCELLANEOUS

As stated earlier, I've always been a super gadget geek, so here are several new experimental tools and technologies being developed to protect people from hurricanes:

1. Hurricane-proof homes: Researchers are working on designing and building hurricane-proof homes using advanced materials and construction techniques. These homes are designed to withstand high winds, heavy rain, and flying debris.
2. Flood-proofing technologies: Researchers are developing flood-proofing technologies, such as flood barriers and water-resistant building materials, to protect against the storm surge and flooding that often accompany hurricanes.
3. Wind-resistant trees: Scientists are experimenting with genetically modifying trees to make them more wind resistant, which could help to prevent damage caused by falling trees during hurricanes.
4. Drones: Drones are being used to gather real-time data on hurricane conditions, which can help emergency responders to assess damage and provide aid more quickly.
5. Early warning systems: Researchers are developing early warning systems that use data from satellites and sensors to detect and track hurricanes more accurately, which can help to improve response times and save lives.

6. Storm surge barriers: Researchers are working on developing storm surge barriers, such as seawalls and floodgates, to protect coastal areas from the devastating effects of storm surges.

7. Artificial intelligence: Artificial intelligence is being used to analyze weather data and predict the path and intensity of hurricanes more accurately, which can help to improve preparedness and response efforts.

Remember that many of these tools and technologies are still in the experimental phase, and it may be some time before they are widely available. In the meantime, it's important to take all available measures to protect yourself and your loved ones from the dangers of hurricanes. Having survived a great many of them, I've found they are to be respected, as their force and power is greater than anything man can do.

8

Twist and Shout

TORNADOES—THE SITUATION

I venture into this arena with huge caution and a certain amount of trepidation. It's not that all the other hazards within this book cannot deliver injury, death, and destruction—they can, and do. But in the case of tornadoes, or twisters, I genuinely feel these ferocious monsters of the plains have a truly fearsome reputation that usurps virtually anything and everything on the surface of the Earth, if only for their destructive potential and acute unpredictability.

Now as it happens, I have personally experienced a near-miss twister about twenty years ago, crossing an adjacent farmer's field next to our home in Buckinghamshire, England. It wasn't a monster, but more a mischievous, late-afternoon rapscallion, known more commonly as a land spout. Either way, the spinning vortex was plain to observe, as was its erratic path and the cacophony of swirling winds as it cut a path across the field and into a nearby woody grove, where it fenced with some equally determined oak trees before finally dissipating.

Mine was a close skirmish with a fitful junior, which brought surprise, curiosity, and fear in equal measures. As we all know, many tornadoes, or their sea-borne equivalents, waterspouts, don't tend to wear diapers and jump straight into jackboots or worse; the biggest of these beasts can carry the force of a thousand-ton truck. But how do these phenomena occur in the first instance, and where?

The "how" is the pure science bit, but allow me to try to keep it simple. Tornado or waterspout activity can only be spawned within a cumulonimbus cloud—the type of cloud that also produces thunderstorms. So, there's a clue straight away. Any ugly black or deep gray skyline, often laced with a backdrop of disarray and likely emanating lightning, can be the start of something big, whether over land or water. However, let me stress that the vast majority of ominous skies and/or thunderstorms do not produce tornadoes or waterspouts; maybe one in five thousand do, some within the heart of other monsters, as in tropical storms. What we require for twisters to happen at all, is firstly the appropriate climate, and then the appropriate local geography, which tends to be vast, open low-lying plains. The appropriate meteorology comes in the form of a strong stream of cold upper air winds and a constant feed of opposing warm, moist air at the surface and in from the sides, which then rises and cools, interconnecting with the higher-level winds and, in turn, creating immense instability and strong updrafts. Pretty soon if the mix is just right, a vortex might form, which can be anything from a few dozen feet across to several miles. Presto, we have a spinner!

The "where" question is the important bit, and you of course want to know whether you are at risk at your location and to what degree. In truth, in a changing world, the answer is that the hit-list map is likely expanding and the thousand-ton truck monsters are roaming somewhat farther than they once did, crashing through historical and geographical boundaries, occurring where previously they didn't. But by far, the majority of the destructive tornadoes form across the plains of the central and southern US, known for good reason as Tornado Alley. It's essentially the battleground between the cold, dry air running southeastward from the Rockies, colliding with the warm, moisture-laden air emanating off the Gulf of Mexico. However, the onward migration of tornadoes has meant regions such as Dixie Alley are now almost equally prone, and it hasn't stopped there—just ask the residents of Mayfield, western Kentucky, who experienced a truly devastating EF4 tornado on December 10, 2021.

EF4? Well, tornadoes are measured by their wind strength and destructive power on the Enhanced Fujita Scale, that's EF1 right up to EF5. Mykel will expand upon the EF scales and what they might mean in terms of wind speeds and your survival requirements.

Meanwhile, did I say December for that catastrophic out-of-order tornado? Yep, and it's unusual, because the majority of tornadoes tend to be spawned during the seasonal crossover months of spring and, to a lesser degree, the fall, when the air masses mix and dance with the devil. So, let's just say that no month can be considered safe when it comes to the potential for tornado formation. The more rigorous and opposing the air masses, the greater potential there is for a tornado to form.

The US might claim the thorny crown for the greatest number of tornadoes per annum, averaging around 1,500, and also the most severe tornadoes in the EF4 and EF5 categories, but here's a list of countries and regions within countries that also endure tornadoes. This list is far from exclusive, and with a warmer and more chaotic climate-changing world, the tornado formation dynamic will almost certainly expand into many other countries and regions in the decades to come:

The Southern Canadian Plains
The UK, in particular England
The Po Valley of Italy
Northeast France
Germany
Poland
Australia
New Zealand
South Africa
China
Japan
Brazil
Northeast Argentina
Uruguay
Southeast Paraguay
Bangladesh
Northeast India
Russia
Ukraine

Waterspouts, on the other hand, tend to form over warm ocean waters, often in tropical or subtropical regions, such as the Florida Keys, Italy, and Greece. However, end-of-summer waterspout activity can occur almost anywhere where the water has warmed sufficiently and we have enough instability in the local atmosphere to spawn those big thunderclouds; there are not that many exclusions.

One little fact to take on board is that a twister of any variety is, in essence, extremely fast-spinning air, which is actually invisible to the eye until it begins to pick up dust, soil, crops, debris, seawater, and other solids or liquids. In its very early stages, you may actually hear it before you see it.

Finally, from me on this one, if your head is not spinning too much, then understand that the mix must be perfect for a tornado or waterspout to form at all, and despite everything that has been said, the formation of these twin terrors in any one location remains a rare event, but one that still merits your acute attention in terms of how to stay safe if one happens upon you.

TORNADOES—THE SURVIVAL

As we continue down the road of looking into the eyeballs of every kind of natural disaster nature can throw at us and finding the best strategies and tactics against it, I like the concept of providing the knowledge of science and facts. This helps you understand the "what," "why," and "how." From there, you can look at some of the options we provide.

Of the things we suggest, some will be applicable, some not. But, where the knowledge first comes into play is in how you apply what we teach to your circumstances now, and maybe even more importantly, what you can do on the fly to adjust your operations to fit the situation when/if it does hit you, directly or indirectly.

SCIENCE AND FACTS

A tornado is a rotating column of air that extends from a thunderstorm cloud to the ground. The basic mechanism that creates a tornado involves the interaction of warm, moist air rising rapidly from the surface with cool, dry air descending from the upper atmosphere. This interaction creates a spinning motion in the atmosphere, which can be intensified by various factors, such as wind shear, convergence of air flows, and temperature differences.

The spinning motion of the air creates a horizontal rotating column, which is then tilted upward by updrafts within the thunderstorm. This tilt causes the rotating column to become vertical and, as it continues to spin, it becomes narrower and faster, creating a funnel-shaped cloud that extends down to the ground. This funnel cloud is the visible part of the tornado.

The intense winds associated with a tornado can cause significant damage, and the path of the tornado can be highly unpredictable. Scientists use a variety of tools to monitor and track tornadoes, including Doppler radar, which can detect the rotation of the air within a storm, and storm chasers, who use a variety of techniques to observe tornadoes up close and gather data about their behavior.

There are several types of tornadoes, each with its own characteristics and formation mechanisms. Here are some of the types:

1. Supercell tornadoes: These are the most common types of tornado and are usually associated with supercell

thunderstorms. They are long-lasting and can be very large and destructive.

2. Multiple vortex tornadoes: These tornadoes have multiple smaller vortices rotating around a common center. They can be highly destructive and unpredictable.

3. Landspout tornadoes: These tornadoes form from the ground up and are not associated with a rotating thunderstorm. They are typically weaker than supercell tornadoes, but can still cause damage.

4. Waterspout tornadoes: These are tornadoes that form over water, such as lakes or oceans. They can move onto land and cause damage if they make landfall.

5. Gustnadoes: These are tornado-like vortices that form along the gust front of a thunderstorm. They are typically weaker than other types of tornadoes.

6. Dust devil tornadoes: These are small, weak tornadoes that form over dry, hot surfaces, such as deserts or parking lots. They are typically not associated with thunderstorms.

Other terms you might hear for tornadoes include *twisters*, *cyclones*, and *spouts*. These terms generally refer to the same phenomenon but may be used in different regions or contexts.

TACTICAL

Tornadoes can have a significant impact on human life, and the severity depends on a variety of factors, such as the strength of the tornado, the location of the person, and the individual's preparedness and response. This is for when you're caught out in it, whether you are unaware, or take a risk to move as it strikes. Having a good understanding of personal impact can help drive better planning and actions. Here are some of the ways that tornadoes can impact humans:

1. Physical injury: Tornadoes can cause serious physical injuries, such as cuts, bruises, broken bones, and head trauma, from flying debris or being thrown around by strong winds.

2. Death: In extreme cases, tornadoes can be deadly, causing multiple fatalities.

3. Emotional trauma: Experiencing a tornado or witnessing its aftermath can be emotionally traumatic for individuals, causing anxiety, fear, and post-traumatic stress disorder.

4. Property damage: Tornadoes can cause significant damage to buildings and infrastructure, leading to costly repairs and displacement for those affected.

5. Disruption of services: Tornadoes can disrupt critical services such as electricity, water, and communication systems, leading to difficulties in accessing basic needs and emergency assistance.

It is important for individuals to have a plan in place for how to respond to tornadoes and other severe weather events, including identifying safe places to shelter, having emergency supplies on hand, and staying informed about the latest weather information.

OPERATIONAL

Let's say you decided to escape to somewhere safe, but got caught in your vehicle. Since the transport is the first step up from tactical/personal, this info is for your operational planning. Being in a vehicle during a tornado is not a safe place to be, as a tornado can cause significant damage to the vehicle and its occupants. Here are some potential outcomes of being in a vehicle during a tornado:

1. The vehicle can be lifted or flipped over: Tornadoes can generate extremely strong winds that are capable of lifting or flipping over vehicles, even if they are moving.

2. Flying debris can cause injury or damage: Tornadoes can pick up and throw debris, such as tree branches or other objects, at high speeds, which can cause significant damage to a vehicle and pose a danger to the occupants.

3. Limited visibility and mobility: Heavy rain and strong

winds can make it difficult to see and maneuver a vehicle, which can make it harder to avoid potential hazards.

4. Being caught in traffic: During severe weather events, roads may become congested with other vehicles, which can limit your ability to move or seek shelter.

If you are in a vehicle and a tornado is approaching, the safest course of action is to exit the vehicle and seek shelter in a sturdy building or underground shelter, if available. If no shelter is available, it is recommended to seek refuge in a low-lying area, such as a ditch or culvert, and cover your head with your hands or a blanket.

For these reasons, it is good to have a radio with a weather station, and you could even use your cell phone, if it still has a signal, to check websites like NOAA for Doppler radar to try and navigate around the weather, but none can accurately predict these things; none ever will. It's like a leaf falling—every single one is different and always will be.

STRATEGIC

Tornadoes can cause significant damage to homes and other buildings, depending on the strength of the tornado and the construction of the building. Tornado damage is much in the same vein as that of the most severe storms, such as hurricanes but, in many cases, can be even more severe. The preventions and safety measures for severe storms, to protect yourself, your home, and property, apply here.

MY STORIES

I have been in a number of tornadoes in my life. Unexpected ones, like while in a movie theatre in Alabama or caught out in the street, and expected ones, where we were ready. I have had a transformer blow up right above my head while standing in a foot of water. I have lived through the night of what sounded like all-out war, and awoke to find the devastation around me in Tennessee a few years back, which was every bit as traumatizing as war. There are no words that can truly prepare you for one that is big, hits hard, and without warning. So, I just stick to my mantra—every day you live, is a day you could die, so make each day count, live well, and do all you can for those you love, while you can. In the end, if fate brings a tornado to you and you survive, you will know in your heart that you did your best, and often enough, that will work, and when it isn't enough, at least it can help you heal, knowing you left nothing to chance that you could control.

MISCELLANEOUS

There are different safe zones for tornadoes, depending on the location and construction of the building or structure. Here are some examples of safe zones that are recommended for tornadoes:

1. Interior room: The safest place to be during a tornado is in a small, interior room, such as a closet or bathroom, on the lowest level of a building. It is recommended to stay away from windows and exterior walls.
2. Basement or storm shelter: A basement or storm shelter is a secure place to seek refuge during a tornado. If a basement is not available, a small interior room on the lowest level of the building can also provide some protection.
3. Reinforced building: Buildings that are reinforced to withstand tornadoes, such as tornado safe rooms or FEMA-approved safe rooms, can provide a high level of protection during a tornado.
4. Mobile home park shelter: Mobile home parks may have designated community shelters or other safe zones that residents can go to during a tornado.

It is important to have a plan in place for where to go during a tornado warning, and to be aware of the safe zones in your area. If you are unsure of where to go, seek guidance from local emergency officials or follow the instructions of a tornado warning alert.

I'm not a big fan of too many numbers when it comes to survival, but it sometimes is necessary to have a basic understanding, as it will help you in a pinch, to make the best decision you can in the heat of the moment. And with a tornado about to drop on you, there are going to be a tense few minutes that will seem like forever. That said, here are the official categories:

Tornadoes are classified based on the damage they cause, using the Enhanced Fujita Scale (EF Scale). The EF Scale is based on estimated wind speeds and the extent of the damage caused by a tornado. Here are the categories of tornado strength according to the EF Scale:

EF0 (65–85 mph or 105–137 km/h): Light damage, such as broken tree branches, damaged siding, and shingles blown off roofs.

EF1 (86–110 mph or 138–177 km/h): Moderate damage, such as broken windows, overturned mobile homes, and roofs partially removed from buildings.

EF2 (111–135 mph or 178–217 km/h): Significant damage, such as roofs torn off well-built homes, cars overturned, and trees uprooted.

EF3 (136–165 mph or 218–266 km/h): Severe damage, such as entire stories of well-built homes destroyed, trains overturned, and heavy cars lifted off the ground.

EF4 (166–200 mph or 267–322 km/h): Devastating damage, such as well-built homes completely destroyed, large buildings heavily damaged, and cars thrown considerable distances.

EF5 (>200 mph or >322 km/h): Incredible damage, such as well-built homes completely leveled, entire neighborhoods destroyed, and cars thrown hundreds of yards.

It's important to remember that all tornadoes are dangerous, regardless of their category. Even a weak tornado can cause significant damage and injuries. It's essential to take all necessary precautions to protect yourself and your family during a tornado. It won't much matter to you when you're in it, the wind speed will be what it will be but the more you understand it, the better you can take countermeasures against it.

Any one of these could be a killer, as even the lightest can cause debris to fly and injure someone. So do your best to be prepared, stay aware, and stay ready to react as best you can. If one of these comes your way, use your head, keep your faith, and keep up the fight.

9

Sand in My Shoes

SAND/DUST—THE SITUATION

When we first suggested this chapter, of all the risks mentioned so far and what's to follow, it appeared to be one of the lesser perils—that is if you don't live in and around the deserts or arid places of the world, or happen upon them on your travels. But, as you will see, sand and dust storms have a rightful place within this book and can be a genuine life-threatening phenomenon, more so if you consider that deserts and arid regions are a growing feature of the globe due to climate change.

So, here we go. A desert is officially classed as any landmass that receives less than twenty-five centimeters (ten inches) of precipitation per annum, or (and I favor this over the former) an area that receives less precipitation than it evaporates. This amounts to a massive one-third of the Earth's land surface and, despite the ice and snow deserts of Antarctica and the Arctic making up a significant part of that, there are numerous expansions of land that are arid or semi-arid and have already become or could easily turn into dustbowls, a process otherwise known as desertification.

However, the real threat here lies in airborne sand or dust, which can travel for hundreds, if not thousands, of miles, and impact the most fertile or most populated parts of the globe. All of which means that sand and dust storms could be a lot closer to you than you may otherwise think.

I've so far used the terms *life-threatening* and *deadly* in talking about what are only grains of sand and specks of dust, hardly, you might think, in the league of extreme temperatures, hurricanes, tornadoes, wildfires, and floods. Wrong! As you will see, the obscure and minute can be every bit as threatening and deadly as the unbounded and immense threats we have covered so far. The threat from sand and dust does not, however, come as a singular entity; no, it arrives in the guise of millions and billions of joined particles.

I'm not a health practitioner in any sense but I know enough and have learned even more to be able to provide you with a list of the known hazards these chaotic storms can create, if they happen upon you.

- Mechanical breakdowns
- Sudden visual impairment
- Keratoconjunctivitis sicca (dry eye disease), possibly leading to long-term blindness
- Asthma
- Allergic reactions and skin ailments
- Asphyxiation
- Heart and cardiovascular impacts
- Dust pneumonia
- Silicosis
- Morbidity
- Cancers
- Interactions with pollution
- Spreading of viral spores, bacteria, toxins and other airborne diseases

There may well be more still lying inside the ugly cupboard, but I hope this is enough for you to appreciate that this particular menace has many barbed teeth and sharp claws, and is worthy of your full attention.

So, let's back up a little and understand what sand and dust are, how such storms occur, and where we are likely to be most at risk. Sand is made up of naturally occurring granular rock particles, with a diameter of between 0.06 and 2.0 millimeters, though I'm guessing there will be plenty of grains within that fall out of those strict measurements, at least the ones that I used to build sandcastles with as a kid. Silt is, however, classed as being finer than sand grains, and grit larger, so now you know.

Sand forms when rocks are weathered or eroded over thousands, if not millions, of years. For the most part, sand is made up of quartz, silicon dioxide, feldspar, mica, natural glass, and other heavy minerals, but fascinatingly it also includes the skeletal remains of billons of organisms of times long past, that is, dead bodies!

The tan color of most sand is the result of iron oxide interaction, which tints the tiny rocks light brown. Black sand, on the other hand, comes from eroded volcanic material such as lava, basalt rocks, and other dark-colored rocks and minerals. There are also several other variations of texture and color, for a plethora of reasons.

Meanwhile, the dust within a dust storm (as opposed to the dust inside your home, largely made up of dead human skin) is generally composed of loose detritus, such as the finest particles of soil, minuscule sand, tiny bits of dead insects, pollen grains, and increasingly microscopic specks of plastic and other man-made materials.

The causes of the storms and, for the purposes of doing your best to avoid the various fallouts, both sand and dust storms follow a similar narrative. Firstly, we must have a source of the material, so the deserts of the world and arid or semi-arid expanses serve the purpose. Next, we require an ultra-dry surface area with no recent precipitation, as in a long-term drought, helpfully intermixed with poor agricultural and/or water practices and an economic/agricultural depression. The catalyst for any dust or sandstorm will then be the lifting, suspension, and transportation of a huge mass of grains/particles off the surface and into the air. Enter the wind!

The wind can arise by various means, in and around tropical storms or thunderstorms, a sudden air mass change usually within a trough or a cold front, or just a simple squeeze of the air pressure gradient, and quite possibly a combination of any of these. Bottom line, we now have an unstable airborne sandstorm or duster on the move, and it's perilous . . . unless you know what to do!

The next questions are where are you most at risk and when? Well, the obvious call is within the proximity of deserts, particularly so in the case of sandstorms. We find the Arabian Peninsula and north Africa at the top of the list, followed by China and central parts of Asia. Australia, parts of North America, South America, and south Africa make up the rest of the most prone zones. Dust storms tend to frequent semi-arid regions adjacent to deserts, where there may be little or no sand, but plenty of baked loose soil, which can easily become a storm's ammunition, given a prolonged drought. Urban areas, too, are their own breeding grounds, given ever-increasing amounts of human waste and scree.

But, and it's an important but, it doesn't end there. As I have already mentioned, sand and dust can be carried hundreds, if not thousands, of miles, even into temperate parts of the US, Europe, and Australia. For example, although nowhere near in the same category as full-on localized storms, upper air southerly winds occasionally carry sand or dust from the Sahara Desert and deposit it over the UK, leaving a thin film of sand or dust on vehicles, window ledges, plant life, crops, and other surfaces. The very same Saharan sand export also gets caught up in the Atlantic trade winds, resulting in valuable nutrients being transported thousands of miles away to the Amazon.

In terms of the most severe debilitating storms and when they might occur, the summer season and the early fall tend to be premium times; it's then that there is an additional amount of energy in the atmosphere, along with an abundance of direct sunshine and excessive heat, greater expanses of dried-up soils, and a likelihood of sudden thunderstorm and/or wind interactions. However, there are no set rules here, and such storms may strike at any time of the year. The following is an appraisal of the most devastating documented sand and dust storms, bearing testament to the importance of the understanding this phenomenon and where the most havoc has been in the thousands of lives lost or ruined, along with some huge economic hits.

- Black Sunday: Texas Panhandle to Oklahoma Panhandle; April 14, 1935, as part of the Dust Bowl events in the US during the 1930s.
- Great Bakersfield Dust Storm: Southern San Joaquin Valley, California; December 19–21, 1977.
- Melbourne Dust Storm: Melbourne, Australia; February 8, 1983.
- Interstate 5 Dust Storm: San Joaquin Valley, California; November 29, 1991.

- Australian Dust Storm: South Australia to New South Wales; September 23, 2009.
- China Dust Storms: China and parts of Vietnam and Thailand; Spring 2010.
- Tehran Haboob Storm: Tehran, Iran; June 2, 2014.
- Indian Dust Storms: North India, May 2–3, 2008.
- North China Super Sandstorm: China, Mongolia, and South Korea; March 2021.
- Iraq Dust Storms: Iraq, April to May 2022.

Finally, from me on this, history may well dictate in respect of the worse-case storm scenarios, but back to those words by J. R. R. Tolkien in the preface: *The world has changed.* The Earth's land surface has gradually become browner and barer due to deforestation, poor agricultural and water practices, bourgeoning human conurbations and, of course, the heat and drought ramifications of global warming. What was once is no longer. The lush, green safety barriers are continually being broken, and the tan and brown deadly grains are not only on the beach and in your shoes, they are in the air and on the move!

SAND/DUST—THE SURVIVAL

For much of this book, Jim has been the information lead as the true weather expert, whereas my contributions are more on what actions you can take before, during, and after calamity strikes, so in short, I'm just more the survival stuff.

Where Jim hits heavy on weather, some other history, trivia, facts, and science, I cover what I think is neat, new, or novel, and which may be relevant for exigency planning. I say all that to say this: Jim nails those things here, as the sand in and of itself may not appear like much, so he proves the point with that adjunct info for this chapter.

That frees me up to just get after courses of action. By the by, as Jim is also a former British weather officer, just like I am a former American special forces officer, we work together to make sure we have you covered for facts and acts!

TACTICAL

Let's get right to it. In this section, Tactical again will apply to the individual countermeasures for sandstorms. If caught out in the open during a sandstorm, it's important to take steps to protect yourself as quickly as possible. Here are some tips:

1. Seek shelter: Look for any nearby shelter, such as a building, car, or rock formation, that can provide protection from the blowing sand and wind.

2. Cover your mouth and nose: Use a mask, bandanna, or any available cloth to cover your mouth and nose. This will help you to avoid inhaling sand and dust, which can cause respiratory problems.

3. Protect your eyes: Wear goggles or any kind of eye protection to keep sand out of your eyes. If you don't have any goggles, use sunglasses, or wrap a scarf or piece of cloth around your head to cover your eyes.

4. Keep low to the ground: Crouch down and cover your head with your arms to protect yourself from flying debris. If you're close to the ground, you'll be less likely to get blown away.

5. Stay put: Once you've found a safe spot, stay there until the sandstorm has passed. Don't attempt to walk or drive through the storm as it can be very dangerous.

Remember, sandstorms can be unpredictable and dangerous, so it's important to take them seriously and take all necessary precautions to protect yourself.

If you're caught in a vehicle during a major sandstorm, most of the same preventions and safety measures that apply for wind and storm events, apply here.

If you happen to be out with an animal—and the following can also be applicable in other severe weather events we've covered—things could get dicey if the animal is spooked and runs. You may lose your animal and never find it again, but also, if your animal is your ride, you may never be able to get out of that area, so take extreme measures to make yourself and your animal as ready as you can with whatever warning time you have. If you're caught in the open during a bad sandstorm with a camel, donkey, or horse, it's important to take steps to protect yourself and your animal in the same way you protect yourself.

Take the same preventions and precautions in maintaining your home and property as you would for all severe weather events, such as making sure your fortress is adequately sealed and locked down, and planning for the care for your family and outreach to your friends and neighbors.

MY STORIES

I've only been in three sandstorms, and in all those cases, they were miserable but not otherwise significant. What I did find interesting, just as Jim said, from wars in Iraq and Afghanistan, we all learned how the daily sand exposure destroys machinery, electronics, gets into everything, causes chafing and cracked teeth, as it coats your food.

PERSONAL

Sandstorms don't hit me these days as I no longer live, work, or serve in any desert areas, but I always found them somewhat exhilarating somehow, as I secretly knew it was going to be messy, but it would be all right. So, I just girded myself the best I could and faced the experience.

MISCELLANEOUS

Sandstorms can create some unusual and even bizarre occurrences. Let's look at some of the strangest things ever recorded as a result of sandstorms:

1. Sand sculptures: During a sandstorm in Saudi Arabia, a group of artists created sand sculptures, including one of a dragon, from the sand that blew into their city.
2. Sand dunes in the ocean: Sandstorms can deposit sand in unexpected places, including the ocean. In 2020, a sandstorm in France created sand dunes in the ocean that were visible from space.

3. Pink snow: In 2019, a sandstorm in Russia combined with snow to create pink snow. The sand particles turned the snow pink, creating a surreal landscape.

4. Blood rain: During a sandstorm in India in 2018, red rain fell from the sky. The sandstorm had combined with thunderstorms, causing the red particles to mix with the rain.

5. Sunken city: In 2018, a sandstorm in China revealed the ruins of a sunken city that had been hidden underwater for decades. The sandstorm had blown away the water and revealed the city's ruins.

6. Disappearing landmarks: Sandstorms can cause landmarks to disappear or become obscured. In 2015, a sandstorm in Egypt covered the Great Pyramids of Giza in sand, making them barely visible.

SECTION V
EARTH/ NATURAL AND ENVIRONMENTAL DISASTERS

The definition of envy is when other planets in our solar system gaze at us and see what we have. The definition of idiocy is to destroy that envy.

10

The Air That I Breathe

POLLUTION—THE SITUATION

Weather and climate are *big*—hopefully you have realized that by now! Together, they arguably have the greatest external impact on our lives, largely determining who we are, what we are about, and what we might do. At least, that is what I have come to believe after many years of studying and quantifying various meteorological impacts and situations. However, from here on, in this penultimate chapter and the one to follow, we are not so much about pure weather and climate as covered within the previous chapters; instead, we are now going to focus on some of the associated factors and hazards that you, your family, and your friends may have to face down, beginning with a hugely impacting one in pollution.

Pollution comes in many forms, and affects us in just as many ways, but by far the majority of foul fallouts are man-made. We have been sadly defecating in our own nest since prehistoric times, though our own discharges have massively expanded in nature and accelerated in the most hellish of ways over the past century or so, and more particularly across the last few decades.

I'm in no way shy or afraid to say that our beautiful planet has been gradually turning into a waste depository and, as the current top sentient beings, we should be thoroughly ashamed. Why, then, has it turned out like this? Well, there are many reasons, some simple and others more complex, but for now, I will offer you five words: *need, greed, selfishness, laziness,* and *shortsightedness.* You can attach whatever political or religious doctrines and inept human actions, all of which have made it into what it is today. The bottom line here is that we have created a cesspit of trouble, nasty foul trouble you would do well to avoid. Sadly, an incredible 6 to 7 million people fail to do so each year, and die due to the various pollutants, ensuring this chapter is every bit, if not more, important than those preceding. Here are the main forms of pollution, and the terrible consequences of each.

AIR

According to the World Health Organization (WHO), air pollution is the contamination of the environment by any chemical, physical, or biological agent that changes the natural characteristics of the atmosphere around us—that is the atmosphere that we live within and breathe, of course.

Motor vehicles, industrial emissions, forest fires, and various household combustion apparatus are the most common sources of air pollution. And, the pollutants of greatest public health concern include particulates, ozone, carbon monoxide, nitrogen dioxide, and sulphur dioxide, in turn causing respiratory diseases, various cancers, and several other deadly or debilitating ailments.

Low- to middle-income countries tend to be most at risk from air pollution, but WHO estimates that a staggering 99 percent of the world's population breathe air that exceeds their guidelines for good health, so there's a pretty fair chance that'll be you. The words *I can't breathe*, in this case, have solid merit. Of course, air quality is closely linked to climate and day-to-day weather, along with global ecosystems, such as the Amazon Forest, often referred to as the lungs of the Earth. And so, given that the burning of fossil fuels is also an acute source of greenhouse emissions, there is a close association between poor air quality and the ramping up of climate change, such that any reduction of combustible pollutants can be considered to be a double victory.

Finally, with concern to air pollution, I can't let this go without mentioning the word *smog*. As a small boy living near Manchester, England, during the 1960s, I remember all too well breathing in the suffocating burning mixture of fog and coal smoke on cold winter mornings, with little or no wind to blow away the acrid fumes. But it was even worse in London during 1952, when the Great Smog is estimated to have killed around ten thousand people. It's a long, painful learning curve, and what I experienced then is still today plaguing other parts of the globe, such as Pakistan, India, and China. Mankind appears to learn very slowly, reluctantly, and painfully, and yet, there are a minority of folks whom, even to this day, when offered a choice, resist any or every move to clean up the air, clinging with religious or political fervor to the dirty and damned status quo, proven time and time again to be a killer. I simply shudder at their blind selfishness.

WATER

Let's be clear from the outset, all forms of pollution eventually make it into our waterways and oceans. They are our handy depositary, and although much of the pollution becomes hidden from our eyes, it is nevertheless there in one form or another, be it debris, rubbish, various chemicals, bacteria, viruses, or parasites. Each of them, and sometimes all of them, together contaminate and make water sources unusable for drinking, cooking, washing, fishing, swimming, and a swath of leisure activities.

Water pollution depletes and destroys ecosystems, and often assists in the growth and spread of poisonous phytoplankton in lakes, ponds, and slower-moving rivers and streams. Once infected, our food chain can become contaminated, and defiled water sources undrinkable. Water-borne diseases can also become rampant with the likes of cholera, diarrhea, typhoid, hepatitis, gastroenteritis, campylobacteriosis, scabies, and worm infections. It truly is a deadly list, and there's more beyond these that I've mentioned.

If we are to look for the world's most polluted waterways, then many paths lead to India, and particularly to the sacred River Ganges, where the incinerated bodies of humans and animals are scattered within the murky, plastic-filled waters, only for pilgrims to then attempt to wash away their sins in the same menacing waters. However, we can very easily go beyond the Ganges in listing other hot spots containing rampant pollution, for example, the mercury-filled Citarum River, east of Jakarta in Indonesia, which is said to claim over fifty thousand lives every year. China's Yellow River is also high on the list of chemical polluted rivers, with water-borne diseases, cancers, and birth defects becoming rife. Europe's bad boy is thought to be the Sarno River, which passes to the south of Naples, with agricultural and industrial discharges the cause of high cases of liver cancer. As for the US, benzene, mercury, and arsenic discharges (to name but a few of the chemicals) make the Mississippi River the most polluted within the country, with both industrial and agricultural waste helping to dilapidate and kill aquatic and plant life alike. There are, of course, many other carriers of wanton waste, including the River Jordan, the Buriganga River in Bangladesh, the Yamuna River in India, Lake Karachay in Russia, the Matanza-Riachuelo River in Argentina, China's Yangtze River, the great River Nile in Africa, and Philippine's Marilao River. You may have your own wretched case on your doorstep, though I sincerely hope not.

LAND

If we have swamped the oceans, seas, and waterways with our pollution, we have also severely degraded the land that also supports us. How, then, have we committed such crimes? Well, as with water exploitation, there are a number of ways. Let's begin with the crime of the centuries, in deforestation and soil erosion. The removal of trees is nothing new, but over the past fifty years or so, the destruction has accelerated exponentially, with woods and forests succumbing to axe, saw, and bulldozer. Sometimes, in hillier or mountainous regions, the removal of trees has unwittingly undermined the land surface itself, resulting in devastating avalanches or heavy rain-associated landslips; you see, weather is seldom out of the equation!

Of course, natural causes have also taken their toll of trees, such as diseases, and animal and pest destruction, not to mention devastation by a massive meteorite (Siberia, June 30, 1908)! However, humankind has been the main slayer, and that also includes soil erosion, caused by the ever-burgeoning construction of roads and buildings, mining, logging, and agricultural practices. Where the soil has been soured and degraded, chemical fertilizers are often introduced in an attempt to compensate for the infertility, but more often than not, that simply causes chemical pollution of the land and any nearby water.

Other forms of land pollution include urbanization, industrialization, sewage treatments, over-crammed landfills, and littering. There is a semi-valid argument that the first five of those have all occurred or been necessary in order to further the progress and causes of mankind, and I would concur to some degree. However, the despoiling of the land in each of these cases has been caused by an amalgamation of growing populations, greed, bad planning, obtuse thinking, corrupt or detrimental legislation, an unwillingness or inability to change, and also not knowing when to stop. As for the bane of littering, there is zero excuse—none, *rien*, zip, zilch. For wanton littering, simply read laziness and indifference.

Radiation: I don't know enough about this particular subject in order to provide you with too much of a brief, save to say that when radiation spills or seeps, the outfalls can, of course, be deadly. I only have to mention Chernobyl or Fukushima to make one realize the perils involved in these cases, due to negligence and natural causes respectively. And then there's Hiroshima, Nagasaki, and remote places around the globe where nuclear detonations have taken place over the ages, since the first one back on July 16, 1945. There are differing air- and ground-retention levels for the different types of radioactivity left behind, though it's as well to know that there are places in this world where the legacies firmly remain to this day.

NOISE, LIGHT, AND PLASTIC

Noise and light pollution may not appear to be in the premier league of contaminating concerns, but what each does is to unveil the extent to which we humans have overburdened ourselves. Both excessive light and over-the-top noise pollution are sources of wasted energy that can degrade environmental quality, causing adverse effects and distress, including mental distress. More particularly, in the case of superfluous light, over-production and overuse is a contributory factor to the unnecessary burning of fossil fuels and climate change. Let's call it what it is—a pure waste. Talking of which, let's bring in the plastic, the same plastic that has been both a blessing and a curse. Yes, we have become a plasticized world in more ways than one, but the waste and debris can be seen all around us in the form of discarded bottles, wrappers, toys, and so much more. If plastic was a living thing, we would call it a gross infestation; as it is, what lies around is a worldwide plague that impacts us all, including wildlife. Burning plastic, meanwhile, creates its own ultra-poisonous hazards, releasing heavy metals, organic pollutants, and toxic chemicals into the air, in turn leading to respiratory and endocrine diseases, as well as various cancers. Despite its obvious usefulness in our daily lives, plastic truly is a Jekyll and Hyde material of our age.

It's generally recognized that the leading pollution centers of the world lie within, or between, the Middle and Far East, rather than in Europe, Africa, or the Americas, but it is as well to know that there are differing pollution hot spots in every continent, aside from perhaps Antarctica, and also within virtually all countries; even the trail to the summit of Mount Everest is literally littered with debris!

In addition, and as you probably know all too well, pollution in its various forms can be very localized. For example, close to where I live and work, we have a small river that eventually feeds into the historic River Thames; sadly, the regional water authority has occasionally seen fit to allow sewage outflows into the river. I could never figure out how it could ever be allowed, but the UK government of 2022–23 passed the legislation to permit it—so much then for their environmental consciousness! I also do my daily exercise across a green field park adjacent to the river, or better still, in the local wood adjacent to the park, where the oxygen levels are higher; anything but running along the traffic- and pollution-stricken London Road, also adjacent to the river. And, then there's that overflowing waste bin we all seem to come across, the one crammed full of empty plastic bottles, bits of leftover food, empty cans, and sweets wrappers, not to mention the ravenous pigeons and the very grateful fat rat!

Pollution is all around us, as are many of the solutions. We put it there, and it's we who must clean up the mess and make right the wrongs; we certainly can't just cross our fingers and hope that the weather's actions over time does it for us—it won't! In the meantime, we have to look to be safe and survive within the somber legacy we have created for ourselves. So, here's how.

POLLUTION—THE SURVIVAL

For me, pollution is real and a real problem. Pollution is important for the survival of the planet and our lives. I'm not going to debate the merit of this topic. If you don't believe it, skip to the next chapter. For now, my metric for courses of action that you can take are within the context of daily, weekly, and monthly. My thinking is, that could take care of a year, or years if need be. Bottom line, there are things you can do for yourself, and others.

TACTICAL

In this section, I just want to give people some ideas of measures they can take if they feel pollution is a threat to their own personal health, or that of their loved ones. Some general tips that may help individuals reduce their exposure to pollution on a daily, weekly, and monthly basis:

DAILY:

1. Check the air quality index (AQI) before going outside, and avoid outdoor activities during peak pollution hours.
2. Use air purifiers and air filters in your home or workplace to improve indoor air quality.
3. Wear a mask when outside, particularly in areas with high levels of pollution.
4. Avoid smoking and exposure to secondhand smoke.
5. Practice good hygiene, such as washing your hands frequently and showering after outdoor activities, to reduce the amount of pollution that you bring into your home.

WEEKLY:

1. Take public transportation, bike, or walk, instead of driving alone to reduce your carbon footprint and decrease air pollution.
2. Eat a healthy and balanced diet with plenty of fruits and vegetables to boost your immune system and reduce the negative impacts of pollution on your body.
3. Use natural cleaning products that don't contain harsh chemicals to reduce indoor air pollution.
4. Support local efforts to reduce pollution, such as participating in community clean-up events or advocating for policy changes.

MONTHLY:

1. Monitor your energy usage and switch to renewable energy sources, like solar or wind power if possible.
2. Use reusable bags, containers, and water bottles to reduce waste and pollution.
3. Choose eco-friendly products and packaging that have a lower environmental impact.
4. Get involved in environmental advocacy, and support policies that address pollution at the local and national levels.

OPERATIONAL

Automobiles will be my standard op-level topic. We all need transportation, so in some way, we all are a part of the problem. There are several things that people can do with their vehicles to reduce pollution. Here are some examples:

1. Drive less: One of the most effective ways to reduce pollution from vehicles is to simply drive less. This can be achieved by using public transportation, carpooling, biking, or walking.
2. Maintain your vehicle: Regular maintenance of your vehicle can help keep it running efficiently and reduce emissions. This includes things like keeping your tires properly inflated, changing the oil and air filters regularly, and ensuring that the engine is running properly.
3. Use eco-friendly products: Choose eco-friendly products such as low VOC (volatile organic compounds) car wash soaps, and cleaning products. Using such products will help reduce the number of harmful chemicals that get washed into our water systems.
4. Use low-emission fuels: Choose low-emission fuels, like ethanol blends or biodiesel. These fuels burn cleaner and produce fewer emissions than traditional gasoline.
5. Avoid idling: Idling your vehicle wastes fuel and produces unnecessary emissions. If you're going to be stopped for more than a minute, it's best to turn off your engine.

6. Drive smoothly and at a steady pace: Aggressive driving, such as rapid acceleration and braking, as well as speeding, can increase emissions. Driving at a steady pace and avoiding sudden stops or accelerations can help reduce pollution.

By taking these steps, individuals can reduce the pollution emitted by their vehicles and help improve air quality in their communities. If everyone did try, it would make a difference.

HOME

Home is our strategic level, as you can't control your neighbors or any municipalities. There are several things people can do to make their homes create less pollution. Here are some examples:

1. Improve ventilation: Improving ventilation can help reduce indoor air pollution. You can open windows and doors to let fresh air in, use exhaust fans in kitchens and bathrooms, and install air purifiers or filters to remove pollutants.

2. Use eco-friendly cleaning products: Choose cleaning products that are labeled as eco-friendly and free of harsh chemicals to reduce indoor air pollution. You can also make your own cleaning products using natural ingredients, like vinegar and baking soda.

3. Properly dispose of hazardous materials: Many household items like batteries, cleaning products, and light bulbs, contain hazardous materials that can pollute the environment. Properly dispose of these items according to local regulations.

4. Use energy-efficient appliances: Energy-efficient appliances use less energy, which means fewer pollutants are emitted from power plants. Look for appliances with Energy Star labels to ensure they are energy-efficient.

5. Low VOC paint: Low VOC paint releases fewer harmful chemicals into the air during and after application, making it a better choice for indoor air quality.

6. Reduce water usage: Reducing water usage can help reduce

the amount of energy required to treat and pump water. You can install low-flow showerheads and faucets, fix leaks, and use a dishwasher or washing machine only when they are full.

7. Reduce waste: Reduce the amount of waste you produce by recycling, composting, and using reusable products. Reducing waste means less pollution created during the production and disposal of products.

8. Natural fiber textiles: Natural fiber textiles like organic cotton and bamboo are grown without harmful pesticides and chemicals and are less likely to release harmful chemicals into the air when washed.

9. LED light bulbs: LED light bulbs use less energy and last longer than traditional incandescent bulbs, which reduces waste and the amount of energy required to produce electricity.

YARD

We don't think about it much, but even our yards can be made better to help reduce pollution. There are several things you can do to make your yard better for the environment. Here are some examples:

1. Plant native species: Planting native plants in your yard can help support local ecosystems and attract native wildlife. Native plants are adapted to local climate conditions and require less water and maintenance than non-native species.

2. Compost: Composting food and yard waste can help reduce the amount of waste sent to landfills, and provide valuable nutrients for your plants.

3. Use natural pest control methods: Avoid using pesticides and herbicides that can harm beneficial insects and animals. Instead, use natural pest control methods like companion planting and handpicking pests.

4. Reduce water usage: Use drought-tolerant plants, collect rainwater, and water your yard in the early morning or late evening to reduce water usage.

5. Use eco-friendly lawn care products: Use eco-friendly lawn care products that are free of harsh chemicals to reduce pollution and protect wildlife.

6. Create a habitat for wildlife: Create a wildlife habitat by providing food, water, and shelter for birds, butterflies, and other wildlife.

7. Install a rain garden: A rain garden can help absorb rainwater and prevent runoff that can carry pollutants into local waterways.

VEHICLE

For this section, I wanted to provide some info about cars that is generally considered most impactful for reducing pollution. The best cars for reducing pollution are typically electric or hybrid cars. These vehicles produce lower levels of emissions compared to traditional gasoline or diesel vehicles. Electric cars do not produce any tailpipe emissions, and hybrid cars combine a gasoline engine with an electric motor to achieve higher fuel efficiency and lower emissions.

Some of the most popular electric cars currently available include the Tesla Model 3, Nissan Leaf, and Chevrolet Bolt. Popular hybrid cars include the Toyota Prius, Honda Insight, and Ford Escape Hybrid. However, it's worth noting that the carbon footprint of electric and hybrid vehicles also depends on how the electricity they use is generated, as well as the manufacturing process and eventual disposal of the batteries.

In addition to electric and hybrid cars, there are also low-emission vehicles that use alternative fuels such as compressed natural gas (CNG), liquefied petroleum gas (LPG), or hydrogen fuel cells. These can offer lower emissions, compared to traditional gasoline or diesel vehicles, but they require specialized refueling infrastructure, which may not be widely available yet.

Now, I am very much an anti-Elon person, so I just want to counterbalance the arguments for electric cars by pointing out some of their negatives, too. While electric cars are generally considered to be better for the environment than traditional gasoline-powered vehicles, there are some arguments against them from an ecological perspective. Here are some of the main ones:

1. Production of batteries: Electric cars rely on lithium-ion batteries to store energy, and the production of these batteries requires a significant amount of energy and resources. Mining for the materials used in batteries can also have environmental impacts.

2. Energy production: While electric cars themselves produce zero emissions, the production of the electricity used to power them may come from non-renewable sources, such as coal or natural gas. This means that electric cars may indirectly contribute to greenhouse gas emissions if the electricity they use is not generated from renewable sources.

3. Battery disposal: At the end of their life, electric car batteries must be disposed of properly. If not, they can potentially harm the environment due to the chemicals they contain. Recycling and disposing of electric car batteries is a process that requires specialized knowledge and equipment, which may not be readily available in all areas.

4. Overall carbon footprint: While electric cars produce no emissions during operation, their overall carbon footprint also includes emissions generated during the production process and from the disposal of batteries. The carbon footprint of an electric car can be lower than that of a gasoline-powered car over its lifetime, but it may not be as significant as initially thought.

It's worth noting, however, that the environmental impact of electric cars can vary depending on factors, such as the source of electricity used to charge them, the type of battery used, and how the batteries are disposed of at the end of their life. Overall, electric cars are still considered to be a more environmentally friendly alternative to gasoline-powered vehicles. Until we have another power source, like magnetic or purely solar powered vehicles with real power to haul weight or go fast, we're in the transition period for humanity that will be as big as the change when we went from animals to motors for powered transport.

MY STORIES

I raised my three sons to have a respect for the land and nature. My eldest, Anthony, takes care of the land for a living. He's very good at it, and I'm very proud of him. My second son, Nicholas, was more into righting wrongs and fixing things—one of my bad habits—and he always went out of his way as a small boy to make sure others picked up their trash whenever he saw folks polluting. Our third son, Gabriel, loves the land, also, and is a master planter. He helps me keep our local waterways clean in our kayaks, or he goes out with his mother, doing group cleanups. We try to practice what we preach, and live it as a way. We are big into recycling and composting.

A small anecdote came to me as I prepared to write this section. As I'm plenty old and crusty, I have forgotten more things than I know, and sometimes, interesting things trigger memories. I recall a time when I was in basic training at Fort Knox, Kentucky, the same place where I was born.

We were just doing a common thing they do with privates—they walk, spread out in a line, and pick up trash, every single cigarette butt or pop-top, nothing that wasn't growing was left. That's why military bases are always so well groomed. It's also a great job we could have citizens do everywhere, for folks on food stamps and other programs; they can help beautify their own areas as a payment back to society if they can't find other work . . . but I digress.

So, while I was picking up trash, and the other troops were complaining, I was picking up everything with a fervor. My drill sergeant walked up on me while I was telling a fellow private how I just could not understand why anyone would want to throw trash down on their homeland—it's as much where we live as our own home. That is when you know you're connected to the land, and it is connected to you. But I didn't grasp the significance of it until I saw my drill sergeant's face. He was perplexed and impressed. It stuck with me. I always see trash. I look at everything for survival potential, but also, it makes me sad. So, I use that as fuel to keep doing better wherever it is within my control to do so.

When I was a kid, I was raised as a Native American, even though it's not my blood. My father was adopted into a tribe called the Mattaponi, and we spent a year playing on the reservation. I recalled the great commercial in the seventies of a Native American seen in the traditional dress of a chief, standing in different locations around America, looking at piles of trash, always in silence, with just a tear rolling down his cheek.

It was effective. It made me not want to pollute. It also made me study about pollution, and a few quotes stuck with me. Here are some of my favorite quotes about fighting pollution:

1. Robert Swan: *The greatest threat to our planet is the belief that someone else will save it.*
2. Native American proverb: *We don't inherit the earth from our ancestors, we borrow it from our children.*
3. Lady Bird Johnson: *The environment is where we all meet; where we all have a mutual interest; it is the one thing all of us share.*
4. Wendell Berry: *The Earth is what we all have in common.*
5. David Attenborough: *We must not only protect the countryside and save it from destruction, we must restore what has been destroyed and salvage the beauty and charm of our cities.*
6. Muhammad Yunus: *The power of human will is unlimited when it is fueled by the passion for what we believe is right.*
7. Native American proverb: *When the last tree is cut down, the last fish eaten, and the last stream poisoned, you will realize that you cannot eat money.* (My personal fave; perhaps that's why money never mattered as much to me as it seemed to for others.)

These quotes remind us of the importance of taking care of the environment and fighting pollution, and encourage us to take action to protect our planet for future generations.

FAMILY

What I want to do here is share some ideas that parents can share with their kids or grandkids to help them learn about, be mindful, and take actions to reduce pollution in general. Teaching kids about reducing pollution can be a great way to help them develop eco-friendly habits and become more environmentally conscious. Here are some things you can teach your kids:

1. Reduce, reuse, and recycle: Teach your kids about the three R's and encourage them to reduce their use of disposable items, reuse items as much as possible, and recycle when appropriate. This can help reduce waste and the amount of pollution produced.

2. Conserve water: Teach your kids about the importance of conserving water by turning off the faucet while brushing teeth, taking shorter showers, and fixing leaks.

3. Walk, bike, or carpool: Encourage your kids to walk or bike to school, or other nearby destinations, or carpool with friends and family members when possible. This can help reduce the amount of pollution produced by vehicles.

4. Plant a garden: Planting a garden can help reduce pollution by improving air quality, reducing soil erosion, and providing habitat for wildlife.

5. Turn off lights and electronics: Encourage your kids to turn off lights and electronics when they're not in use. This can help reduce energy consumption and the amount of pollution produced by power plants.

6. Buy eco-friendly products: Teach your kids about eco-friendly products and encourage them to choose products made from sustainable materials or that have minimal packaging.

7. Participate in community cleanup events: Participating in

community cleanup events can help teach your kids about the importance of taking care of the environment and reducing pollution in their local community.

By teaching your kids about these and other eco-friendly habits, you can help them develop a greater appreciation for the environment and become more responsible stewards of the planet.

FRIENDS AND NEIGHBORS

There are some fun things you can do with friends and neighbors to fight against pollution. Try some of these good things that groups can do to fight pollution:

1. Organize community cleanup events: Organize cleanup events to remove litter and pollutants from local parks, waterways, and other public spaces. This can help improve the health of the environment and bring community members together to work toward a common goal.

2. Advocate for change: Advocate for policies and regulations that promote sustainability and reduce pollution. Attend local government meetings and speak out in support of environmental initiatives.

3. Promote eco-friendly practices: Educate others about eco-friendly practices and encourage them to adopt sustainable behaviors like reducing waste, conserving energy and water, and using public transportation.

4. Plant trees and native plants: Planting trees and native plants can help remove pollutants from the air, prevent soil erosion, and support local ecosystems.

5. Support local businesses that prioritize sustainability: Support local businesses that prioritize sustainability, eco-friendliness, and reducing pollution.

6. Volunteer with local environmental organizations: Volunteer with local environmental organizations to support their efforts to reduce pollution and promote sustainability.

7. Start a conversation: Engage your neighbors in a conversation about the importance of reducing pollution and how it can benefit both their health and the environment. Be respectful and listen to their concerns.

8. Share information: Share information about local air and water quality, and the impact of pollution on human health and the environment. Provide resources that can help them learn more about the issue.

9. Lead by example: Set an example by adopting eco-friendly practices and showing your neighbors how easy it can be to reduce pollution. For example, you can install solar panels on your roof, compost your food waste, or use public transportation instead of driving alone.

By making these changes, you can make your home, yard, and vehicle more sustainable and eco-friendly spaces that support local ecosystems and protect the environment.

INTRAPERSONAL

I have a constant sensitivity meter to my immediate surroundings that serves me for survival in the woods and in wars, but it also makes me ever-aware of my presence in any space and my impact on it. I try not to leave any trace other than footprints, and sometimes, I try not to leave those. As a way to keep me aligned with how I, and we, impact our environment and how much love I have for unspoiled nature, I focus on these things. Living a clean life regarding pollution involves adopting eco-friendly habits and making conscious choices to reduce your impact on the environment. Here are a couple of additional steps you can take to live a clean life regarding pollution:

1. Eat a plant-based diet: Eating a plant-based diet can reduce your carbon footprint and help reduce pollution caused by animal agriculture.

2. Support renewable energy: Choose renewable energy sources, such as solar or wind power, when possible, and advocate for policies that promote renewable energy.

By making these changes, you can reduce your impact on the environment and live a cleaner, more sustainable life. It's important to remember that every small change you make can have a positive impact on the environment and, collectively, our efforts can help create a cleaner and healthier planet for future generations.

INTERPERSONAL

Most of the time, if you see someone throwing trash down, you just let it go and look at them with a bit of sadness and disgust while you try to understand what made them that way or why they are that way. Sometimes, if someone is polluting, and you just can't let it be and want to convince them to stop, there are several approaches you can take. Here are some strategies:

1. Educate them: Often, people are not aware of the impact their actions have on the environment. Educate them about the harmful effects of pollution, and explain how their behavior is contributing to the problem.

2. Appeal to their values: Try to understand what motivates the person, and frame your argument in a way that appeals to their values. For example, if the person cares about the health of their family, explain how pollution can harm their loved ones.

3. Provide alternatives: Offer the person alternatives that are less harmful to the environment. For example, if they are using plastic bags, suggest reusable bags instead.

4. Show them the impact: Sometimes, people need to see the impact of pollution firsthand to understand its severity. Take them to polluted areas or show them photos or videos of polluted environments.

5. Collaborate: If the person is unwilling to change their behavior, try collaborating with them to find a solution that works for both parties. For example, if they are unwilling to stop using their car, suggest carpooling or using public transportation instead.

6. Lead by example: Show the person how you are reducing your own impact on the environment, and how it has benefited you. Lead by example and inspire them to make changes too.

It's important to approach the conversation with empathy and respect, and to avoid being confrontational or judgmental. By using these strategies, you can encourage the person to take steps to reduce their impact on the environment and help create a cleaner, healthier planet for all.

MISCELLANEOUS

Before I leave you on this, as I know many may see this chapter as rubbish, I just want to share some interesting facts for you to dwell on. Without further ado, here are some unique or trivia facts about pollution that people may find interesting:

1. The Great Pacific Garbage Patch, a concentration of plastic debris in the North Pacific Ocean, is estimated to be twice the size of Texas.
2. The world's first air pollution law was passed in 1306 in England. The law prohibited the burning of soft coal, which produced smoke and sulfur dioxide.
3. Every year, approximately 8 million metric tons of plastic waste enters the world's oceans, which is equivalent to one garbage truck of plastic being dumped into the ocean every minute.
4. Indoor air pollution can be up to ten times worse than outdoor air pollution, due to factors such as poor ventilation, mold, and household chemicals.
5. In the early 1900s, pollution from coal burning was so bad in London that it caused a thick fog, known as a pea-souper, that was so thick it caused accidents and deaths.
6. The world's largest source of pollution is the burning of fossil fuels for energy, such as coal, oil, and natural gas.

7. Plastic waste can take up to a thousand years to decompose in the environment, and when it does, it releases toxic chemicals that can harm wildlife and ecosystems.

These facts demonstrate the severity of pollution and the urgent need to take action to reduce our impact on the environment. Anyone who does not believe we are having a negative impact on our world by polluting it, is either not looking, or they're knowingly lying, in my opinion.

11

It's the End of the World as We Know It

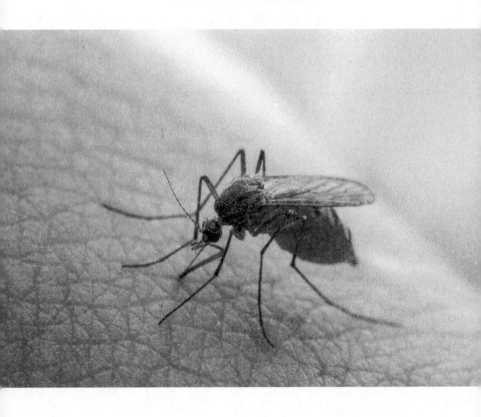

FROM EARTHQUAKES TO PLAGUES—
THE SITUATION

And so, we arrive at the beginning of the end, or is it the end of the beginning?

In this, the final chapter, I'm going to let you in on a not-very-well-kept secret; it's more of an expectation than a secret, but in the end, and by one means or another, mankind *will* be wiped off the face of the Earth—or maybe the Earth will be wiped off the face of mankind?

Thankfully—hopefully—neither scenario is imminent, and there is no set date that you need to worry about just yet. All being well and at a long stretch, we are 5 billion-odd years away from being burnt to smithereens by our dying expanding sun, which conveniently leaves plenty of time for us to explore and provide antidotes for a plethora of other perils that may present themselves in the interim, such as the white-hot topic of climate change.

By now, you should have a good grasp of how weather and a changing climate might impact us in one form or another; some being pure weather events, others such as dust storms and pollution being associated risks. As I have stated earlier, in certain circumstances, the perils may combine and be a cause for double trouble, examples being hurricanes spawning tornadoes; blizzards combining with a deep freeze; windstorms issuing in floods; severe heat or lightning resulting in wildfires. Every one of these scenarios, and many more besides, can then go on to produce other unlooked for impacts well beyond the realms of honest weather, but all of them potentially inspired by a global climate on the change. Then, there are the rock-based geological impacts, and it's there we will begin. And so, for the final time, let's explore in an attempt to play safe!

EARTHQUAKES, TSUNAMIS, AND VOLCANIC ERUPTIONS

Strictly speaking, none of these terrible phenomena are meteorological events; that's despite meteorological agencies throughout the world tending to be responsible for the monitoring and warnings of these bad boys. This arena, and the associated sciences, concern the Earth's physical structure and substance and the various processes that act upon it. In that, we have to delve deep into the dark places of the world, to a place where only the balrogs roam (*Lord of the Rings* fans, you are with me, right?).

The link between the three cataclysms is a close one, in that they are joined by the hip with one event often directly leading to another. In essence, we are talking here about the movement of rock, solid or molten, and the consequences, small- or large-scale. There are no meteorological factors involved here beyond what might occur in the wake of such devastating events, which can, and often do, turn into acute perils of their own. Instead, our focus is on the movement of the tectonic plates, huge sections of the Earth's crust and upper mantle with a thickness of around a hundred kilometers, together known as the lithosphere. There are fifteen main plates and another fifteen minor ones, all of which have been on the move for around 3 billion years, so nothing new there then. However, most earthquakes, tsunamis, and volcanic activity revolve around the ongoing movement of these plates. When the stress on the edge of a plate overcomes the immense friction that is taking place, a sudden release of energy occurs, which travels through the Earth's crust, resulting in an earth tremor or quake which, in turn, within oceanic areas can manifest itself into tsunamis. When one plate is sliding or subducting below another, a line of volcanoes forms along the overriding plate. The regions of the world where the tectonic actions are most apparent are well known, and for most people who live in such areas, it is a constant fact of life, a little like living with erratic heart palpitations that one day may explode. If you are visiting a region where the earth may literally move or explode into life, then it's as well to know it and here's why:

If we take each of the perils in turn, the main initial threat to life from each is as follows:

- Earthquakes: Collapsing buildings, falling objects and flying objects.

- Tsunamis: Drownings and traumatic injuries.
- Volcanoes: Burning, asphyxiation, and traumatic injuries.

There are many other grim reapers within each, not forgetting other extreme perils that follow such catastrophes such as landslides, mudflows, famine, and disease—all of which we will turn to shortly. When these giant terrors of the Earth occur, the death and devastation can be truly immense. The Indian Ocean tsunamis of December 2004 killed an estimated 250,000 people. While back in January 1556, in Shansi, China, some 830,000 lost their lives to a huge earthquake. More recently, the Turkish earthquakes of March 2023 sadly took almost 60,000 lives, making it the deadliest natural disaster in modern history. Meanwhile, the eruption of Mount Tambora, Indonesia, in 1815, led to the deaths of an estimated 92,000 people, the majority not directly from the volcanic activity itself, but from starvation in the aftermath due to the land and nearby ocean being devastated by ash and lava.

Let's be clear here, we often get warnings of imminent hurricanes, floods, tornadoes, and other extreme weather events, but in the case of these terrible beasts, scientists have yet to be able to see clearly into the crystal ball in enough time to allow evacuations. In some cases, in certain parts of the world, we are literally living on the brink.

LANDSLIDES

This is when the earth really does move for you, except in such cases it's definitely not where you want to be! A landslide, or landslip, is defined as the downslope movement of a mass of rock, soil, mud, and any accompanying debris. The four main types of earth movement are labeled as being falls, slides, flows, and topples, each having a different reason to occur. Landslides or slips can be initiated in slopes already on the verge of movement by heavy or sustained rainfall, snow melt, changes in ground water levels, erosion by rivers or streams, earthquakes or tremors, volcanic activity, disturbance by human activities, or any combination of those factors.

Of all the catalysts, the penetration and movement of water is considered to be the greatest. It's simple physics, really, but heavily saturated soil and rock can be at risk from succumbing to gravity, thereby excessive rainfall or snowmelt are prime dangers where there is any degree of fragility, at, or just below, the surface. It perhaps goes without saying, the removal of trees on hillside slopes due to human activity (and in that, I would mention the words greed and stupidity) can, and often does, lead to the undermining of the hills or mountainsides to the detriment of whoever happens to live or be passing below. Very often, the result can be a devastating mudslide—a thick, soup-like flow that moves with great speed and force and can literally smother unwary communities in a blanket of suffocating death within seconds.

Landslides can vary immensely in size and destructive potential, as we have already discovered, but perhaps the most infamous and visually captivating landslide occurred in 1980, when the whole side of Mount Helens collapsed during volcanic activity. The most-deadly landslide occurred in the city of Huaraz, Peru, in December 1941, when a flow of water, mud, and debris—care of a catastrophic glacial lake overflow—accounted for up to five thousand unfortunate souls. In my own memory, and it sticks in my mind even though I was a very young boy at the time, the coal slag landslide of October 1966 in Aberfan, Wales, was particularly horrific. One hundred forty-four people, including 116 children in the local school situated within the mountain valley, succumbed to a cascade of unstable coal waste and mud. After numerous heart-rending campaigns and inquiries, the totally avoidable tragedy was eventually pinned on executives at The National Coal Board, despite insurers at the time shamelessly squirming, claiming it was an act of God. Yeah, that one normally fails the court test for sound reasons—and did!

ANIMAL AND INSECT INVASIONS, PLAGUES, AND NEW DISEASES

It might only take a small nudge in a particular environment to change, or even destroy, a particular organism or even a whole ecosystem. It's happened on countless occasions in the past and will continue to happen into the future. Of course, the bigger the impact on the environment, the greater the change or destruction, none more massive than the Great Dying, or the Permian-Triassic extinction of 250 million years ago, thought to have been caused by the warming of the Earth that gradually wiped out around 85 percent of living species over an estimated 10-million-year period. I guess we can call that a cataclysmic climate change in action!

What I am driving at here is the flapping of a butterfly's wings (more commonly known as the butterfly effect), coined in the 1960s by fellow meteorologist Edward Lorenz. What that actually means is that miniscule occurrences on one side of the globe could, in the end, translate (according to computer modeling) into much more massive and decidedly impacting events on the other side of the globe or, for that matter, just next door. In reality, it's very hard to say or to predict such outcomes, but certain words tend to crop up in the dialogue, such as sensitivity, randomness, disorder, irregularities, and the big one that we know is alive and well in the meteorological world, namely chaos. Of course, left to their own devices, things (that's everything in the universe) become less structured, so there tends to be a natural element of disorder in the physical world, even though in the relatively short term, physics strives for balance and neutrality.

So, let's take that chaos and disorder, perhaps ushered in by an unnatural warming of an ocean or a great forest by a couple of degrees, or maybe a semi-arid land turning into a desert, or permafrost in Greenland or Siberia giving up the ghost and millions of tons of methane in the process. I could give many other simple or exotic examples, and there will be plenty more that we don't yet know but will materialize over the course of time. Any one of them could result in manifest changes to ecosystems and potentially everything that lies within them. Of course, all forms of life strive to live long and prosper, but when acute changes occur and disorder rules, it really is backs to the wall, and it's every human, animal, insect, plant, fungi, bacterium virus, or thing, to him or herself! It's called survival of the species, and some species are far more long-lived and adapted to managing chaos than we currently are.

Now, I don't really wish to go down the road of *The Last of Us* for those who have seen it, nor *Outbreak, 28 Days Later, World War Z, I Am Legend, Contagion, Bug, Them,* and a whole host of apocalyptic films, series, and books that have centered on the rise of weird stuff we would prefer not to ever happen, so let's think instead about something that did happen, namely COVID-19. Was it first created in a Chinese laboratory as some would claim, or was it a mutation that moved from bats, birds, pigs, or another animal into humans? At the time of writing this book, that question still hangs in the air, but it might be equally fair to say that particular virus may have mutated because of environmental changes; remember, in whatever circumstance it occurred within, it was simply attempting to adapt, survive, and proliferate.

Of course, it may not be the most minute of life forms that cross into untrodden boundaries, or metamorphosize into something we might not be able to handle. There are hundreds, thousands, and billions of creatures of every shape and size that climate change or epic weather might either wipe out altogether, increase exponentially, or merit a rapid change in their DNA. I believe we can expect to witness all three scenarios in the decades to come, quite possibly on countless occasions. Those scary fiction stories of viruses and insect invasions may, in the end, be not too far off the mark, so let's just hope that technology and good sense come to the rescue and it doesn't spell the end for us!

DROUGHT AND FAMINE, BATTLES, AND WARS

Given the previous subjects, and now these, you might have thought the Four Horsemen of the Apocalypse had already descended upon us. Well, they haven't, and it's our job to provide a little help and advice in order to keep it that way. So, where are we going here with these particular horrors?

Any severe event, but particularly events that are manifest and impact whole countries or regions, can create political or economic instability. Weather impacts are very high up on that impact list, and with climate change upping the stakes across the continents of Africa and Asia in particular, there are already skirmishes happening concerning precious natural resources, with lack of water being the main one. According to the Pacific Institute, we are already approaching thirteen hundred known skirmishes or minor battles over water alone. Iraq, Syria, Turkey, Egypt, Somalia, and Ethiopia to name but a few countries, have all found themselves in water scraps of one kind or another, and for sure, those numbers and the size and scale of such conflicts, will grow as the world's climate zones become more unstable.

The bottom line here is that water, being the most valuable commodity on Earth, also largely governs the plants and crops that can be grown and harvested. So, this is as basic as it gets in terms of the water that we need to drink and the food that we need to eat. Take away or restrict the basics of life from a community, a region, or a country, and there's every chance that the apple cart will tip over into something unsavory.

Drought and hunger are sure drivers of anxiety, fear, and desolation, and that acute desperation can, and will, lead to mass human migrations, rather like the massive herds of wildebeest on the African plains, ignoring the many dangers as they forever hoof it in search of their next available sustenance. Except this: in the human world, there are strict borders and distinct ownerships not to be crossed, and what is theirs is theirs, if you get my drift. That, my friend, is where the root causes end, and the bloody strife begins.

Now, we are not including this section to alarm and make you head for the hills with your tent and backpack, rather it's included to alert you as to how weather and climate change (whether natural or not) can eventually spiral out of control to the point of emotional and physical warfare, with the epicenters always likely to be in the drier and poorer parts of the globe—at least for now.

Finally, here's a personal one for your back pocket, and hopefully one never to be used within your lifetime. When it comes to it, weather is big in war and battles (I know I was once that sailor, with a big anti-aircraft gun within a war zone!). Throughout history, dozens of encounters, large and small, have been won or lost on the vagaries of the weather, whether it be major snow and freeze in the World War II Battle of Stalingrad (the USSR was saved); the destruction of the Spanish Armada, in 1588, by stormy conditions rolling in off the North Atlantic (Britain was saved); and last but not least, in August 1776, heavy rain and fog provided the necessary cover for George Washington and his men to retreat and escape the British at the Battle of Long Island (the US was saved). Now, just imagine for a moment how history could have turned out if the weather hadn't played its part!

MENTAL HEALTH

I am going to finish up with this one, given everything we have mentioned so far comes down to how we might handle any of it from a mental point of view. I'm sure you know what I'm getting at when I say that, in my mind, the weather is *the* prime catalyst for how we might feel and how we might fare on a day-to-day basis. In short, weather can be a mood-maker or a mood-breaker. Allow me to explain.

When we awake and check out with our beady eyes what is going on outside weather-wise, it leaves an immediate imprint on our minds. For most of us, a blue sky and bright sunshine inspire the chemical serotonin to be produced within our brains. Serotonin is essentially a feel-good chemical that regulates mood, emotion, appetite, and digestion. Around thirty minutes exposure per day is considered to be enough to help lift the spirits and enhance our positivity.

All of this almost certainly goes back to our prehistoric ancestry, when fine weather allowed for the most basic of activities in terms of hunting and gathering food. Add in some greenery in the form of trees and plants, and the colors sky blue, yellow, and green are considered by clever scientists that study such things as the prime colors in helping people to overcome many forms of clinical depression, or simply to bounce back from an adversity.

Of course, too much of anything can be as detrimental as too little. Waking up to a relentlessly blazing hot sun every day, as in some of the deserts of the world, can channel the mind in a negative way and cause day-to-day distress, such that when the one day of cloud and rain finally arrives, the serotonin and its sister chemical dopamine can then do somersaults for the opposite reason!

In terms of excesses, climate change and extreme weather events can be a cause for post-traumatic stress disorder, depression, and high levels of anxiety.

Extreme heat or extreme cold, relentlessly high humidity, incessant rain, fog, and smog, or even something as banal as leaden gray skies, can all have a detrimental effect on our mental state, not to mention having to suffer the losses and/or pick up the pieces of a powerful destructive event.

Let's be clear, whether it's one of those windy days that seem to make kids cavort excitedly around the school playground, or a thunderstorm that leaves people gripping the sides of their bed in fear at night, or that first joyous snowflake of the season, or that helpless, languid feeling when the air is crammed full of water vapor, and a dozen other situations, for good or ill, the weather is a huge player in the mental health stakes—and one we all need to seriously take onboard.

FROM EARTHQUAKES TO PLAGUES— THE SURVIVAL

For this section, I'm throwing out all the previous categories, as you've learned enough on all the different ways you can do things for all the different aspects of your life, in all the different types of major physical calamities of hot, cold, wet, dry, fast, and dirty.

Now, we're going to look at the downright catastrophic things that, while they're not the literal end of the world, it sure will feel like it to anyone in that part of the world. There will likely be the kind of death tolls that shake everyone on earth when you speak of tens of thousands or millions of human deaths in an instant.

There are a few ways to look at these. If you're right there, you die right away. Most think that is good; if you have to go, why not fast? But if you survive, there are two different schools of thought. Some will say surviving is just a slow death. Maybe so. But most would say, that's living. However, the degree of difficulty will often be commensurate to the distance from the epicenter of the disaster—closer to the incident, the worse it is.

With that, let's look at each of the earth-shattering events and give you some ideas for what you may be able to do to better navigate survival after one of these catastrophes. No one could ever prepare for these, nor know what they will find or have to use, or to do. That is why I love survival—it is simply creative problem-solving within a dynamic environment with really, really high stakes! Keep that sense of humor; often, it can make the difference.

EARTHQUAKES

I was in Russian language school when the big one hit San Francisco, with a 6.9 magnitude on October 17, 1989. I'd never been in an earthquake before, and at first, I wasn't sure what it was. I remembered to get out of the house, and when I did, I thought it was normal California stuff. But when the locals who grew up there were all running around screaming it's the end of the world, I gathered this was *not* normal. If you're caught in an earthquake, here are some courses of action you can consider.

INSIDE:

1. Drop to the ground and take cover under a sturdy desk or table. Hold on to it until the shaking stops.
2. Stay away from windows, shelves, and other items that could fall or break.
3. If you can't get under a desk or table, crouch down in an interior corner of the building and cover your head and neck with your arms.
4. If you're in bed, stay there, and cover your head with a pillow.
5. Don't try to run outside or use elevators.
6. If you're in a high-rise building, stay away from windows and exterior walls. Move to a lower floor if possible.

OUTSIDE:

1. Move away from buildings, trees, and power lines.
2. If you're in a car, pull over to a clear area and stay in the car with your seatbelt on until the shaking stops.

3. If you're near a tall building, move away from it to avoid falling debris.
4. If you're near the ocean, move to higher ground immediately to avoid a possible tsunami.
5. If you're in a mountainous area, be aware of the possibility of landslides, and move to a safe location.

Remember to stay calm, assess your surroundings, and follow these guidelines to keep yourself safe during an earthquake.

There are several technologies that can help detect earthquakes, including:

1. Seismometers: These are instruments that measure ground motion caused by seismic waves. They are the most common tool used to detect earthquakes.
2. GPS sensors: These sensors can detect the displacement of the ground caused by an earthquake. They can also provide information about the location and magnitude of the earthquake.
3. Infrasound sensors: These sensors detect low-frequency sound waves that are produced by earthquakes. They can detect earthquakes that are too small to be picked up by seismometers.
4. Satellite imagery: Satellites can detect changes in the Earth's surface caused by an earthquake. This technology is particularly useful for detecting large earthquakes.
5. Early warning systems: These systems use a combination of sensors and algorithms to detect earthquakes and provide warnings before the shaking reaches populated areas. They can help reduce the risk of injury and damage from earthquakes.

There are some consumer technologies that people can buy to help detect earthquakes. These include:

1. Smartphone apps: There are many earthquake alert apps available for both iOS and Android devices. These apps

use real-time data from seismometers to provide users with alerts when an earthquake occurs.

2. Personal seismometers: Some companies sell small, portable seismometers that individuals can use to monitor seismic activity in their area. These devices typically connect to a smartphone or computer to provide real-time data.

3. Home earthquake alarms: Some companies also sell devices that can be installed in homes to detect earthquakes. These alarms can be set to sound an alarm when they detect seismic activity.

It's important to note that these consumer technologies may not be as reliable as professional-grade equipment used by scientists and emergency responders. If you live in an area prone to earthquakes, it's important to have an emergency plan in place and to follow the instructions of local authorities in the event of an earthquake.

TSUNAMIS
If you're near a tsunami-prone area, there are several signs that a tsunami may be about to occur. These include:

1. Strong earthquake: Most tsunamis are caused by underwater earthquakes, so if you feel a strong earthquake that lasts for more than twenty seconds, it could be a sign that a tsunami is on its way.

2. Abnormal ocean behavior: If you see the ocean receding suddenly, exposing large areas of the seafloor, or if you notice a rapid rise or fall in sea level, it could be a sign that a tsunami is approaching.

3. Unusual animal behavior: If you notice birds or marine animals behaving abnormally, such as moving to higher ground or acting agitated, it could be a sign that they sense a tsunami is coming.

4. Loud roar: If you hear a loud roar that sounds like an oncoming train, it could be a sign that a tsunami is approaching.

It's important to note that not all earthquakes or abnormal ocean behavior will result in a tsunami. However, if you notice any of these signs, it's important to take them seriously and follow the instructions of local authorities for evacuation or sheltering in place.

If you are in an area where a tsunami is expected, there are several things you can do to increase your chances of survival:

1. Evacuate to higher ground: If you are in a low-lying area near the coast, move to higher ground as quickly as possible. Tsunamis can travel inland for several miles, so it's important to get to high ground that is at least a hundred feet above sea level, or two miles inland.

2. Follow official instructions: If you receive a tsunami warning, follow the instructions of local authorities. They will provide guidance on evacuation routes and safe areas.

3. Avoid the coast: Stay away from beaches, harbors, and other coastal areas until local authorities say it is safe to return.

4. Seek shelter: If you cannot evacuate to higher ground, seek shelter in a sturdy building on higher ground. Stay inside until the tsunami has passed.

5. Be prepared: Have an emergency kit with food, water, and other supplies, as well as a first aid kit and any necessary medications. Keep important documents in a waterproof container.

6. Stay informed: Monitor local news and weather reports for updates on the tsunami and follow the instructions of local authorities.

Remember, tsunamis can be very dangerous and unpredictable, so it's important to take them seriously and take appropriate action to protect yourself and your family.

There are several new technologies that have been developed to help people survive a tsunami. In addition to advanced warning systems and cell phone apps, these include:

1. Tsunami-resistant buildings: Engineers are developing

buildings that are designed to withstand the force of a tsunami. These buildings typically have reinforced concrete or steel structures and are built on higher ground or elevated platforms.

2. Vertical evacuation structures: In some areas where there is limited high ground, vertical evacuation structures have been developed. These structures are multistory buildings that can provide shelter for people during a tsunami. They are designed to withstand the force of the waves and have emergency supplies and equipment stored on higher floors.

3. Autonomous underwater vehicles: Researchers are developing autonomous underwater vehicles that can quickly survey the ocean floor after a tsunami to assess damage and locate survivors. These vehicles can operate in conditions that are too dangerous for humans and can provide valuable information for search and rescue efforts.

It's important to note that these technologies should be used in combination with other preparedness measures, such as emergency planning, evacuation drills, and education on tsunami risks and warning signs.

VOLCANIC ERUPTIONS

Ruth and I filmed one of our best shows ever on *Man, Woman, Wild* when we got the chance to film on Montserrat. It was a capitol city completely abandoned overnight and half-covered in volcanic lava and ash. What it meant to me was two things. One, I could teach the first real urban survival, and our lives were at risk the entire time, as the iconoclastic flow that threatened us could come with no warning and kill us in under a minute.

But it also meant I had more to work with than on any other show. So, I made a primitive radio and a high-speed fire piston. It's worth a watch, on Discovery, Amazon, and other online platforms.

So, let's continue to review the signs, the actions, and the protections.

There are several signs that can indicate a volcano is about to erupt. Here are some of the common signs:

1. Increased seismic activity: Volcanic activity often causes increased seismic activity, including tremors, earthquakes, and ground deformation.

2. Gas emissions: As magma rises to the surface, it releases gases such as sulfur dioxide, carbon dioxide, and water vapor. Changes in the amount and composition of these gases can signal an impending eruption

3. Changes in surface features: Bulges, cracks, and other changes in the surface of a volcano can be an indication that magma is moving beneath the surface.

4. Changes in volcanic activity: Changes in the frequency, intensity, or type of volcanic activity can also signal an impending eruption.

5. An increase in temperature: As magma moves closer to the surface, it can heat up the surrounding rocks and ground, leading to an increase in temperature.

It's important to note that while these signs can be useful for predicting volcanic eruptions, they are not foolproof. Volcanic eruptions can be unpredictable, and it's important to stay alert and follow the advice of local authorities if you live near an active volcano.

If you find yourself caught in the middle of a volcanic eruption, here are some important things to keep in mind:

1. Follow the instructions of local authorities: If you're in an area that's at risk of a volcanic eruption, stay informed by following updates from local authorities and emergency services. They will provide you with the latest information and advice on how to stay safe.

2. Evacuate if necessary: If you're instructed to evacuate, do so immediately. Don't delay or wait to see what happens.

Follow the recommended evacuation route and head to a designated safe zone or shelter.

3. Protect yourself from falling ash: Cover your nose and mouth with a damp cloth or mask to protect yourself from inhaling ash. Wear protective clothing such as long-sleeved shirts, pants, and sturdy shoes. Use goggles to protect your eyes from ash and debris.

4. Avoid low-lying areas: Stay away from low-lying areas, as they are at risk of flash floods and mudflows (lahars) caused by heavy rainfall mixing with volcanic ash and debris.

5. Stay indoors if possible: If you're not able to evacuate, stay indoors, and keep all doors and windows closed. Use a towel or other cloth to seal any gaps around doors or windows to keep ash out.

6. Listen for warnings: Listen for warnings of additional eruptions or changes in the situation. Be prepared to evacuate immediately if necessary.

Remember, volcanic eruptions can be extremely dangerous, and it's important to take all necessary precautions to protect yourself and your loved ones.

There are various technologies that can help detect and protect people from volcanic eruptions. Here are some examples:

1. Volcano monitoring systems: These systems use various techniques to monitor volcanic activity, such as seismic sensors, GPS devices, and gas sensors. By analyzing the data collected from these devices, scientists can detect changes in volcanic activity and issue warnings to people who may be at risk.

2. Volcanic ash detectors: These devices use advanced technology to detect volcanic ash in the air. They can help aviation authorities track the spread of ash clouds, which can be hazardous to aircraft.

3. Lahar warning systems: Lahars are mudflows that can occur

during a volcanic eruption when ash and other volcanic debris mix with water. Lahar warning systems use sensors and alarms to alert people to the danger of a potential lahar.

4. Geothermal power plants: Geothermal power plants use heat from volcanoes to generate electricity. While not specifically designed for volcanic monitoring, these plants can be used to detect changes in the temperature of the surrounding area, which can be an indication of volcanic activity.

5. Protective gear: Various types of protective gear can be used to shield people from the harmful effects of volcanic eruptions, such as ash masks, goggles, and heavy-duty clothing.

It's important to note that while these technologies can be useful for detecting and protecting people from volcanic eruptions, they are not foolproof. It's always best to follow the advice of local authorities and emergency services in the event of a volcanic eruption.

LANDSLIDES

There are several indicators that a landslide may occur. Here are some of the common signs:

1. Ground movement: If you notice the ground around you moving, sliding, or sloping, it could be a sign that a landslide is imminent.

2. Cracking or bulging in the ground: If you notice new cracks or bulges in the ground, it could be a sign that the soil is shifting and a landslide may occur.

3. Changes in the landscape: If you notice sudden changes in the landscape, such as changes in the river flow, changes in the vegetation or trees leaning, it could indicate that a landslide is about to occur.

4. Unusual sounds: If you hear unusual sounds, such as cracking or rumbling noises, it could be a sign of an impending landslide.

5. Water accumulation: If you notice water accumulation on the ground or on the slopes, it could be a sign of an

impending landslide. Water can weaken the soil, making it more susceptible to sliding.

It's important to note that while these signs can be useful for predicting landslides, they are not always present before a landslide occurs. Landslides can be unpredictable, and it's important to stay alert and follow the advice of local authorities if you live in an area that is at risk of landslides.

If you find yourself caught in a landslide, here are some important things to keep in mind:

1. Try to get out of the path of the landslide: If possible, try to move out of the path of the landslide. Move to higher ground, if available, and stay away from the landslide's edges.
2. Protect your head: Cover your head and face with your hands or with any available object to protect yourself from falling debris.
3. If you're inside, stay inside: If you're inside a building, stay there and take cover under a sturdy piece of furniture, such as a table or desk, until the landslide is over.
4. Listen for warnings: Listen for warnings of additional landslides or changes in the situation. Be prepared to evacuate immediately if necessary.
5. Be aware of secondary hazards: Landslides can cause secondary hazards such as flash floods, debris flows, or rockfalls. Stay alert to the potential for these hazards and be ready to evacuate if necessary.
6. Call for help: If you are injured or need assistance, call for help as soon as it is safe to do so. Use your phone or other communication device to call emergency services or to signal for help.

Remember, landslides can be extremely dangerous, and it's important to take all necessary precautions to protect yourself and your loved ones.

There are various technologies that can help detect and protect people from landslides. In addition to landslide monitoring and early warning systems, here are some examples:

1. Landslide barriers: These barriers are designed to prevent landslides from reaching populated areas. They can be made of various materials such as concrete, steel, or geotextiles, and can be installed on slopes to divert or contain landslides.

2. Geotechnical engineering techniques: Geotechnical engineering techniques, such as slope stabilization, soil nailing, and retaining walls, can be used to reinforce unstable slopes and prevent landslides from occurring.

3. Landslide-resistant buildings: Buildings in landslide-prone areas can be designed to withstand the impact of landslides. Features such as reinforced foundations, flexible connections, and building setback from the slope, can increase the resilience of buildings to landslides.

It's important to note that while these technologies can be useful for detecting and protecting people from landslides, they are not foolproof. It's always best to follow the advice of local authorities and emergency services in the event of a landslide.

MUDSLIDES

Mudslides, also known as debris flows, are a type of fast-moving landslide that can be triggered by heavy rain, snowmelt, or other factors. Here are some of the common indicators that a mudslide may occur:

1. Heavy rainfall: Mudslides are often triggered by heavy rainfall, especially if the soil is already saturated with water.

2. Steep slopes: Mudslides are more likely to occur on steep slopes, where the soil is less stable and more prone to movement.

3. Soil erosion: If you notice soil erosion on slopes or hillsides, it could be a sign that a mudslide is possible.

4. Cracks or bulges in the ground: If you notice new cracks or bulges in the ground, it could be a sign that the soil is shifting and a mudslide may occur.

5. Unusual sounds: If you hear unusual sounds, such as cracking or rumbling noises, it could be a sign of an impending mudslide.

6. Vegetation changes: Changes in the vegetation, such as the sudden appearance of dead or dying plants, can indicate that the soil is becoming unstable and a mudslide may occur.

It's important to note that while these signs can be useful for predicting mudslides, they are not always present before a mudslide occurs. Mudslides can be unpredictable, and it's important to stay alert and follow the advice of local authorities if you live in an area that is at risk of mudslides.

If you find yourself caught in a mudslide, much of the same preventions and safety measures for landslides apply; the same for the various technologies for early warnings, barriers, and engineering, etc.

AVALANCHES

Avalanches are simply landslides of snow. They can be triggered by a variety of factors, including weather conditions, snowpack stability, and slope angle. Here are some of the common signs that an avalanche may occur:

1. Recent snowfall: If there has been a recent, heavy snowfall, it can increase the risk of an avalanche.

2. Changes in temperature: Rapid changes in temperature, such as a sudden warming trend, can cause the snowpack to become unstable.

3. Wind: High winds can transport snow and cause the formation of unstable snowdrifts, increasing the risk of an avalanche.

4. Cracking or collapsing snow: If you hear a cracking or collapsing sound while walking on the snowpack, it could be a sign that the snow is unstable and an avalanche may occur.

5. Visible signs of instability: If you notice signs of instability in the snowpack, such as cracks or fractures in the snow, it could indicate that an avalanche is possible.
6. Recent avalanche activity: If there has been recent avalanche activity in the area, it can increase the risk of another avalanche occurring.

It's important to note that while these signs can be useful for predicting avalanches, they are not foolproof. Avalanches can be unpredictable, and it's important to stay alert and follow the advice of local authorities if you plan to travel in avalanche-prone areas.

If you find yourself caught in an avalanche, here are some important things to keep in mind (also note that many of these actions can apply to other forces that have the potential to bury you):

1. Try to get out of the path of the avalanche: If possible, try to move out of the path of the avalanche. Move to the side of the avalanche, or try to grab onto a solid object like a tree or rock to stop yourself from being carried away.
2. Try to stay on the surface: If you are caught in the avalanche, try to stay on the surface of the snow. Make swimming motions with your arms and legs to keep yourself afloat.
3. Create an air pocket: If you get buried in the snow, try to create an air pocket in front of your face with one arm while keeping the other arm extended above your head.
4. Try to create space around your mouth: If you're wearing a helmet or goggles, try to create space around your mouth so that you can breathe more easily.
5. Use rescue equipment: If you have avalanche rescue equipment, such as a transceiver, probe, or shovel, use them to try to locate and dig out anyone who may be buried.
6. Stay calm: Stay calm and conserve your energy. Try to keep your head clear and focused, and conserve your energy until help arrives.

Remember, avalanches can be extremely dangerous, and it's important to take all necessary precautions to avoid being caught in one. If you plan to travel in avalanche-prone areas, make sure to check the avalanche forecast, carry appropriate equipment, and travel with a partner or group.

There are various technologies that can help detect and protect people from avalanches. Here are some examples:

1. Avalanche beacons or transceivers: These devices emit a signal that can be used to locate someone who has been buried in an avalanche. They are commonly worn by skiers, snowboarders, and other outdoor enthusiasts who travel in avalanche-prone areas.

2. Airbag systems: These systems consist of a backpack with an airbag that can be deployed in the event of an avalanche. The airbag helps to keep the person on the surface of the snow, making it easier to locate and rescue them.

3. Avalanche probes: These long, collapsible poles can be used to search for someone who has been buried in an avalanche. They are commonly used in conjunction with avalanche beacons.

4. Remote sensing technologies: Remote sensing technologies, such as satellites, radar systems, and drones, can be used to detect changes in the snowpack that may indicate an increased risk of avalanches.

5. Avalanche control measures: Avalanche control measures, such as snow fences, explosives, and ski cutting, can be used to trigger controlled avalanches in order to reduce the risk of uncontrolled avalanches.

It's important to note that while these technologies can be useful for detecting and protecting people from avalanches, they are not foolproof. It's always best to follow the advice of local authorities and experts in the event of an avalanche.

ANIMAL AND INSECT INVASIONS

There are various indicators that a pest or insect invasion may be imminent. Here are some of the common signs:

1. Presence of the pest: If you notice the presence of the pest or insect in your home or surrounding areas, it is a clear indication that an invasion may occur.

2. Damage to plants or crops: If you notice damage to your plants or crops, such as holes in leaves, wilting, or stunted growth, it could be a sign that pests or insects are feeding on them.

3. Increased sightings of the pest's natural predators: If you notice an increase in the number of natural predators of the pest or insect, it could be an indication that a population surge is about to happen.

4. Changes in climate or weather: Changes in climate or weather conditions, such as increased rainfall or temperature, can create favorable conditions for the growth and reproduction of pests and insects.

5. Unusual behavior of the pest: If you notice unusual behavior of the pest or insect, such as a change in feeding habits, or if they become more active than usual, it could be a sign that they are preparing for a population surge.

It's important to note that while these signs can be useful for predicting pest or insect invasions, they are not always present before an invasion occurs. It's best to take preventive measures such as proper sanitation, removal of potential breeding sites, and use of pest control methods to avoid pest or insect invasions.

If you find yourself caught in the middle of a dangerous swarm of rodents or locusts, here are some important things to keep in mind:

1. Seek shelter: If possible, seek shelter in a building or vehicle. Close all windows, doors, and vents to prevent the rodents or locusts from entering.

2. Cover your face and eyes: Cover your face and eyes with a mask or cloth to prevent the rodents or locusts from getting into your eyes, nose, or mouth.

3. Avoid inhaling dust: Rodents and locusts can kick up a lot of dust, which can irritate your lungs and make it difficult to breathe. If you are outside, try to cover your mouth and nose with a mask or cloth to avoid inhaling dust.

4. Stay calm: Try to remain calm and avoid panicking. Rodents and locusts are more likely to attack if they sense fear or panic.

5. Use insect repellent: If you have insect repellent, use it to repel the locusts or rodents. However, be careful not to spray it directly on yourself, as it can be harmful if ingested or inhaled.

6. Call for help: If you are in a dangerous situation, call for help as soon as possible. Use your phone or other communication device to call emergency services or to signal for help.

Remember, swarms of rodents or locusts can be extremely dangerous, and it's important to take all necessary precautions to protect yourself and your loved ones.

There are various technologies that can be used to protect individuals from dangerous plagues of pests. Here are some examples:

1. Pesticides: Pesticides are chemical substances that can be used to kill or control pests. They can be applied to crops, homes, or other areas to prevent or control pest infestations.

2. Traps: Traps can be used to catch and control pests, such as rodents or insects. There are various types of traps available, such as snap traps, glue traps, and live traps.

3. Netting: Netting can be used to protect crops from pest infestations. It can be placed over plants to prevent pests from reaching them.

4. Ultrasonic devices: Ultrasonic devices emit high-frequency sound waves that are intended to repel pests, such as rodents

or insects. These devices can be used in homes or other areas to deter pests from entering.

5. Biological controls: Biological controls involve the use of natural predators, such as birds or insects, to control pest populations. This can be an effective and environmentally friendly way to control pests.

It's important to note that while these technologies can be useful for controlling pests, they should be used responsibly and in accordance with safety guidelines. It's also important to take preventive measures, such as proper sanitation and removal of potential breeding sites, to avoid pest infestations.

PLAGUES AND NEW DISEASES

There are various indicators that a new disease or plague may be emerging. Here are some of the common signs:

1. Unusual symptoms: If people are experiencing unusual symptoms, such as fever, cough, or respiratory distress, it could be an indication of a new disease or plague.

2. Increase in cases: If there is a sudden increase in the number of cases of a particular disease or illness, it could be a sign of an emerging plague.

3. Geographic spread: If a disease or illness is spreading rapidly across different regions or countries, it could be an indication of an emerging plague.

4. Unexplained deaths: If there are unexplained deaths occurring in a particular area or community, it could be a sign of a new disease or plague.

5. Animal outbreaks: If there are outbreaks of illness or disease occurring in animals, such as livestock or wildlife, it could be a sign that a disease is spreading to humans.

6. Lack of effective treatments: If there are no effective treatments or vaccines available for a particular disease, it could be an indication that a new disease is emerging.

It's important to note that while these signs can be useful for detecting emerging diseases or plagues, they are not always present before an outbreak occurs. It's best to stay informed about disease outbreaks and take preventive measures, such as practicing good hygiene, avoiding contact with sick individuals, and seeking medical attention if you experience symptoms.

If you find yourself caught in the middle of a plague or new disease outbreak, here are some important things to keep in mind:

1. Follow official guidance: Follow official guidance from health authorities, such as the World Health Organization (WHO) or the Centers for Disease Control and Prevention (CDC). Stay informed about the situation and follow any instructions or recommendations given by health authorities.
2. Practice good hygiene: Practice good hygiene by washing your hands frequently with soap and water, covering your mouth and nose when coughing or sneezing, and avoiding touching your face.
3. Avoid contact with sick individuals: Avoid contact with individuals who are sick or showing symptoms of the disease. If you are sick, stay home and avoid contact with others to prevent the spread of the disease.
4. Wear protective gear: If necessary, wear protective gear such as a mask, gloves, or goggles to protect yourself from exposure to the disease.
5. Seek medical attention: If you experience symptoms of the disease, seek medical attention immediately. Call ahead to your healthcare provider or local hospital to inform them of your symptoms and to receive instructions on how to proceed.
6. Stay home if possible: If the disease is highly contagious, stay home and avoid contact with others to prevent the spread of the disease.

Remember, it's important to take all necessary precautions to protect yourself and others from the spread of a new disease or plague.

Stay informed, follow official guidance, and seek medical attention if necessary.

There are various technologies that can help detect and protect people from plagues or new diseases. Here are some examples:

1. Diagnostic tools: Diagnostic tools, such as PCR tests or rapid antigen tests, can be used to detect the presence of a disease in an individual. These tests can be useful for early detection and containment of the disease.
2. Contact tracing apps: Contact tracing apps can be used to track and trace individuals who have been in close contact with someone who has the disease. This can help contain the spread of the disease by identifying and isolating those who may have been exposed.
3. Telemedicine: Telemedicine allows patients to consult with healthcare providers remotely, reducing the risk of exposure to the disease in a healthcare setting. This can be particularly useful during a pandemic or outbreak.
4. Personal protective equipment (PPE): PPE, such as masks, gloves, and gowns, can be used to protect individuals from exposure to the disease. Proper use of PPE can help reduce the risk of infection.
5. Air filtration systems: High-efficiency particulate air (HEPA) filters can be used to remove airborne particles, including viruses, from the air. This can be particularly useful in healthcare settings or areas with poor air quality.
6. Vaccines: Vaccines can provide immunity against a disease, reducing the risk of infection and transmission. Vaccines are particularly useful for preventing the spread of diseases during a pandemic or outbreak.

It's important to note that while these technologies can be useful for detecting and protecting people from plagues or new diseases, they should be used responsibly and in accordance with safety guidelines. It's also important to take preventive measures, such as practicing good hygiene, avoiding contact with sick individuals, and seeking medical attention if you experience symptoms.

DROUGHT AND FAMINE

There are various indicators that drought and famine may be imminent. Here are some of the common signs:

1. Lack of rainfall: If there is a prolonged period of low rainfall, it can lead to drought conditions, which can increase the risk of famine.
2. Dry soil: If the soil is dry and cracked, it can be a sign of drought conditions.
3. Reduced crop yields: If crop yields are lower than usual, it can be a sign of drought conditions. This can lead to food shortages and an increased risk of famine.
4. Water shortages: If there is a shortage of water, it can be a sign of drought conditions. This can lead to a lack of water for drinking, irrigation, and other essential needs.
5. Increase in food prices: If there is an increase in the price of food, it can be a sign of a shortage of food due to drought or other factors.
6. Conflict and displacement: Drought and famine can lead to conflict and displacement as people struggle to access food and water.

It's important to note that while these signs can be useful for predicting drought and famine, they are not always present before a drought or famine occurs. It's best to take preventive measures, such as water conservation, sustainable agriculture, and emergency food and water supplies, to avoid drought and famine.

If you find yourself caught in a drought or famine, here are some important things to keep in mind:

1. Seek assistance: Seek assistance from local authorities, relief organizations, or humanitarian groups. They can provide

food, water, and other essential supplies to help you and your family survive during the drought or famine.

2. Conserve water: Conserve water by using it wisely and avoiding waste. You can collect rainwater or use greywater for non-potable purposes such as irrigation or flushing toilets.

3. Practice sustainable agriculture: Practice sustainable agriculture by using drought-resistant crops and methods that conserve water and nutrients. This can help to reduce the impact of the drought or famine on crops and food supplies.

4. Use food and water wisely: Use food and water wisely by rationing and prioritizing essential needs. Avoid wasting food and water and try to share resources with others who are also affected.

5. Stay informed: Stay informed about the situation and any updates from local authorities or relief organizations. This can help you to stay prepared and informed about any assistance that may be available.

Remember, drought and famine can be extremely challenging and it's important to take all necessary precautions to protect yourself and your loved ones. Seek assistance, conserve water, practice sustainable agriculture, and use food and water wisely.

While there are various technologies that can be used to predict drought and famine, many of these technologies are typically used by governments, international organizations, or research institutions. However, there are some technologies that a private family could use to help them monitor and prepare for a drought or famine. Here are some examples:

1. Soil moisture sensors: Soil moisture sensors can be used to measure the moisture content of the soil. This information can be used to determine whether the soil is dry and if irrigation is necessary.

2. Rainwater collection systems: Rainwater collection systems

can be used to collect and store rainwater for non-potable purposes, such as irrigation.

3. Drought-tolerant plants: Planting drought-tolerant plants can help to conserve water and reduce the impact of drought on crops and gardens.

4. Solar-powered water pumps: Solar-powered water pumps can be used to pump water from a well or other source for irrigation purposes.

5. Weather monitoring systems: Weather monitoring systems can be used to track weather patterns and predict when a drought or other extreme weather conditions may occur.

It's important to note that these technologies can be useful for monitoring and preparing for drought and famine, but they may not provide a complete solution. It's also important to take preventive measures, such as water conservation, sustainable agriculture, and emergency food and water supplies, to avoid drought and famine.

BATTLES AND WARS

There are various indicators that there may be a risk of war or battle. Here are some of the common signs:

1. Increase in military activity: If there is an increase in military activity, such as troop movements or military exercises, it could be an indication that tensions are rising and there may be a risk of conflict.

2. Political instability: Political instability, such as government instability, protests, or civil unrest, can increase the risk of conflict.

3. Disputes over territory or resources: Disputes over territory, resources, or other issues can lead to tensions and increase the risk of conflict.

4. Arms buildup: If there is an increase in the production and acquisition of weapons, it could be a sign that a country or group is preparing for conflict.

5. Threats and rhetoric: Threats and hostile rhetoric from leaders or groups can increase tensions and raise the risk of conflict.

6. Economic sanctions: Economic sanctions or other measures taken by one country against another can increase tensions and raise the risk of conflict.

It's important to note that while these signs can be useful for predicting the risk of war or battle, they are not always present before conflict occurs. It's best to stay informed about political and social events in your region and to take precautions to stay safe in case of conflict.

If you find yourself caught in the middle of a war or battle, here are some important things to keep in mind to protect yourself and your family:

1. Stay informed: Stay informed about the situation and any updates from local authorities or international organizations. Listen to official guidance and follow any instructions or recommendations given by authorities.

2. Find a safe place: Find a safe place to take shelter, such as a basement or interior room. Avoid areas with windows or external walls, as they are more vulnerable to shrapnel and bullets.

3. Avoid crowds: Avoid large crowds or public gatherings, as they can be a target for attacks.

4. Keep a low profile: Keep a low profile and avoid drawing attention to yourself. Stay quiet and avoid making noise or turning on lights, as this can attract attention.

5. Stock up on supplies: Stock up on essential supplies, such as food, water, and medical supplies. Have a supply of cash on hand in case banks and ATMs are unavailable.

6. Stay connected: Stay connected with loved ones and keep a charged mobile phone with you at all times. Have a designated meeting place in case you get separated.

7. Plan for an escape: Have a plan for escape in case the situation worsens. Identify possible escape routes and have a backup plan in case your first plan is blocked.

Remember, war and battle can be extremely dangerous and it's important to take all necessary precautions to protect yourself and your loved ones. Stay informed, find a safe place, avoid crowds, keep a low profile, stock up on supplies, stay connected, and plan for an escape.

While there are no technologies that can guarantee avoiding being caught in a war or battle, there are some technologies that can help individuals stay informed and prepared. Here are some examples:

1. Mobile applications: Mobile applications, such as emergency notification apps, can provide real-time updates on security situations and help individuals avoid dangerous areas.

2. Satellite imagery: Satellite imagery can be used to monitor military activity and track movements of troops or vehicles. This information can help individuals avoid areas of conflict.

3. Social media: Social media can be used to stay informed about security situations and connect with local communities for advice and guidance.

4. Personal safety devices: Personal safety devices, such as emergency alarms or tracking devices, can help individuals alert authorities or loved ones in case of an emergency.

5. Virtual private networks (VPNs): VPNs can be used to access the internet securely and avoid censorship or surveillance in areas of conflict.

It's important to note that these technologies may have limitations and cannot guarantee safety in all situations. It's also important to stay informed about the situation and take necessary precautions, such as avoiding dangerous areas, having emergency plans, and seeking advice from local authorities or trusted sources.

Epilogue

JIM

And finally, we are here, at the end of all things weather and climate, impacts of all shapes and sizes, along with some of the other stuff we have tagged on, just in case any of them happen upon you. It's been a journey, and now we are at the end of this particular road.

The majority of hazardous occurrences we have described within this book remain a rarity, and most of us alive today will only experience a small handful of such adversities in our lifetimes. However, the world we both work within merits that it is far better to be advised and safe than it is to be heedless of the possibilities and eventually sorry.

Nobody really knows what the future has in store, and precisely how and when our changing climate might impact by way of perilous weather and/or the other forms of hazards mentioned within. But, for when the Earth does have a need to weep and show a tear, we sincerely hope that you will now be a little more prepared than you were previously. Thank you for being with us on this journey. Stay safe and always take the weather with you!

MYKEL

My friends, if you've read this, then you've already taken the important first steps. But I'd encourage you not to stop there. Make sure you've done all you can do, before you rest. If time or dime delays you, that's okay; make a plan and a schedule, and try to keep to it. We are never fully prepared, as there is always more you can do. So, do what you can, as you can, and if you know you're doing your best, then you can persevere with the confidence that you're giving it all, and often enough, that is what will see you through.

For me personally, I have long known and accepted that we're all going to die, someday, some way, so there's no point fearing it. Just be your best self, live your best life, strive to help others, and to make the world a bit better wherever you are and however you can. And here's a parting thought, that may or may not offend—no offense is intended.

But I hold to the belief that there is a great spirit of goodness in the universe, and most of us call that God, by some words or others. I choose to use the word God. And I believe in science. To me, it is nothing more than us understanding at a macro or micro level, how God makes things happen. With that in mind, we now know from the Webb Telescope that there are an infinite number of galaxies beyond all we ever knew or even thought, which makes sense to me. How? If the diversity of life on Earth is so much greater than anything we could have conceived just a few decades ago, with all manner of incredible life forms from mycelium to microbes, deep ocean to pitch-black cave life, deep jungles to barren deserts, then why can't there be millions more planets and life forms?

To me, it all reinforces the greatness of the great spirit of goodness. That intelligence is so much higher than ours—why wouldn't we want that in a deity? So, for all my friends who have issues with all the science, please don't. The ability to understand that science is as much a gift to mankind as all the other blessings Mother Earth and nature give us. Be thankful, learn, and never stop.

I also believe that for every problem we face, there is a solution. I have faith in our teachers to give the next generation the best tools we can, and that those kids will find a way. Be it on this planet, or the next.

So, just like we can generally plan, as a human, to live a certain amount of years, like seventy to ninety, unless something happens, we can likely plan on another billion years of homesteading on Earth. And, in that time, just like we have injuries and illnesses, we will have good and bad times as a human race and as an Earth planet species. But, if something catastrophic happens, we might not be able to change what happens, or even the outcome, but we can always choose how we face it.

And finally . . .

Never quit. Always think. Do your best. Survive with honor.

And may God be with us all.

Appendices

YOU'LL NEVER WALK ALONE

APPENDIX I

Philosophy, Psychology, and Mentality

MANAGING HOPE AND EXPECTATIONS

As a master martial artist, longtime guerrilla warfare expert and practitioner of extreme survival globally, I have found a few key things that help me get through tough times. Some are general, others, quite specific.

Firstly, survival is an expedition in the toleration of misery. Bugs, heat, cold, hunger, thirst, fatigue, illness, injury, threats, and danger are all part of that harsh reality's paradigm. It does no good to complain. So instead, we focus of doing.

Do whatever you can, when you can, as you can, to make things better for yourself and others. This is simple, but it is a mighty powerful secret sauce. Here's why. When you stay busy, you vanquish fear by facing your potential threats and making plans to counter them. You fight despair, because you've taken control of your situation the best you can in that time and space, even if you only get control of yourself—that is more mastery than most attain in their normal lives. We count these as little victories, and we hang on to them in our dark times.

Likewise, the more you do, the more you see, the more you find, the more you learn, the more you can exploit or turn to your advantage or avoid for your preservation. In short, doing makes things better. There's an ancient Chinese quote I have always taught: *Expectation is the source of all unhappiness.*

Think about that—it's very true. If you expect to be saved, and it doesn't happen, that can have a catastrophic impact on morale. However, if you decide to *hope* for rescue, but do not expect it and take all subsequent actions and decisions accordingly, then *if* you're rescued, it's a win; if not, you're doing all you can from the get go, to ensure you rescue yourself and your loved ones or, at least, preserve yourselves as long as you can. Again, hope should always be your guide, but expect nothing. Hope is not a plan—words and deeds that help, are a good plan.

MANAGING PEOPLE IN TRIALS AND TRIBULATIONS

Next, is the interpersonal skills of dealing with others. Some will be scared, others angry, others lost, others present. You will have to navigate these by contemplating what could go wrong, what dilemmas you could face, what decisions you'd make. Do not take the easy button here.

In special forces officer training, they throw many dilemmas at us to evaluate our core character and judgment. We have to discern and detangle the complexities of the "human terrain" in order to find the best way for the group. No single solution set exists, as each human and moment will always be unique to that time and space.

For example, what if you want to go/evacuate, but the majority wants to stay. You're worried about a storm coming, and others are worried that leaving the downed plane will lead to not being found. There are a multitude of potential pros and cons; do you have a radio, is anyone injured, etc. These are the kinds of hard decisions that may need to be made. Literally, your life will depend on it, and theirs. So, make time to consider these, too.

Then there is the intrapersonal, or the thoughts within you. You must not only fight the elements, and potentially others, but you will have to fight your own battles within, as well. You may be sick or hurt and have a low tolerance for other's opinions, you may be unfit and not able to make a journey, you may have your own fears and doubts, prejudices and preconceptions, good and bad, that you'll have to face while you're hungry, thirsty, and/or tired—it is key to know your core personality, know how you are in hard times, and counter those aspects you don't care for.

Much will depend on the dynamics of your family and/or group. The key to remember is everyone will be scared. The raw power of nature is primal, and one must master their own fear to face it and remain calm. Fear is a deadly virus that does spread like wildfire. It can only be put in check by calmer minds prevailing. So, face it, overcome it, and be one of the calming solutions, not one of the panicking problem creators.

Think through what drastic measures may be needed. Plan for them, make peace with them, and have triggers at decision points that are preplanned so they tell you what to do immediately when something happens, instead of catching you off guard, causing a pause or delay in critical thinking and timely decision making.

GROUP AND REGROUP

I have been in more than a few disasters in combat and nature. In war, I learned that panic can spread as fast as lightning. I learned the only counterbalance is calm. Slow, deliberate words and deeds. Set the example and keep folks busy. They say idle hands do the devil's work for a reason. When folks aren't busy focusing on improving their situation, mentally they are busy thinking of all the different ways things can go wrong. So, here are some tips to help as a leader. During an extreme event or situation, it is crucial to prioritize the mental well-being of everyone involved. Here are some strategies to help keep people mentally together during hardship:

1. Stay calm and positive: As a leader or organizer, it's important to stay calm and positive to help maintain a sense of stability and security for those around you. Encourage everyone to remain optimistic and remind them that help is on the way.

2. Communicate clearly: During an extreme event, it can become difficult or impossible, so it's important to establish a communication plan ahead of time. Make sure everyone

knows what to do if they get separated or if there's an emergency.

3. Keep everyone warm and fed: Ensure everyone has access to adequate and appropriate clothing, blankets, and food. This can help alleviate some of the physical discomfort associated with the extreme weather conditions and can help people feel more secure.

4. Encourage group activities: If possible, encourage group activities like games or storytelling to help maintain a sense of community and connection. This can help to distract from the stress and anxiety of the situation.

5. Practice self-care: Encourage everyone to practice self-care. This could involve taking breaks, engaging in relaxation techniques like meditation or breathing exercises, and seeking support from others.

6. Stay connected: If possible, stay connected with loved ones and the outside world. This can help to alleviate feelings of isolation and can provide reassurance that help is on the way.

FRIENDS AND NEIGHBORS

I'm perfectly happy to be alone, but my family loves people and I love them, so like many folks, I have neighbors. While I respect their privacy and definitely want them to respect mine, when bad things happen, good people do good things.

You're going to be in it together, whether you like it or not. Be sure to look out for them, check in on them, if you don't see them about afterwards, go looking for them. If they seem absent, report it so the authorities can get them on their radar, as everyone will be striving for accountability immediately after, so anyone lost, hiding, or stuck, has the best chance of being found alive.

Even in our currently very politically divided country, we are all still Americans, and most people are good, even if we disagree. In the worst of times, we have an opportunity to show the best of our humanity, or the worst. You'll have to live with your choices, and one day, you'll have to answer for them, good or bad, this life, or the next—always conduct yourself with honor. Make that decision now, and it will serve you well, for the most part.

So, here are some things you can do to be a bit more neighborly, or not! There are several things you can do to help your neighbors during an adverse event.

1. Check on your neighbors: If you know that your neighbors are elderly or have a medical condition, consider checking on them to make sure they are safe and offer assistance if needed.
2. Offer shelter: If your neighbors are outside and don't have access to shelter, offer to let them come inside your home or garage.
3. Share information: If you know that your neighbors are not aware of the risks or the precautions they should take, share information with them on how to stay safe.
4. Check outdoor furniture, equipment, etc.: If your neighbors have outdoor equipment such as a TV antenna or satellite dish, check to make sure they are properly grounded to reduce the risk of lightning damage; in stormy conditions, check to make sure furniture, grills, toys, etc., are secured.
5. Report hazards: If you notice any hazards in your neighborhood, such as downed power lines or tree limbs, report them to the appropriate authorities as soon as possible.

ACKNOWLEDGE, COMFORT, ENGAGE

1. Stay calm yourself: It's important to remain calm and composed when helping someone who is panicking. This can help to keep the situation from escalating and make it easier for the person to listen to you.
2. Validate their feelings: Acknowledge the person's fears and concerns. Let them know that it's okay to feel scared, but that there are things you can do to stay safe.
3. Listen to them: Allow the person to express their feelings

and concerns without interruption. Listening can help them to feel heard and understood, which can be calming.

4. Provide reassurance: Provide reassurance that you are prepared and that you have a plan in place to stay safe. Let them know that you are there for them and that you will do everything you can to protect them.

5. Focus on the present: Focus on the present moment and the things you can control. Encourage the person to take deep breaths and focus on their breathing.

6. Offer practical solutions: Offer practical solutions, such as finding a safe place to shelter or engaging in calming activities like reading or listening to music.

7. Use humor: Using humor can sometimes help to defuse a tense situation and make the person feel more relaxed.

Remember that everyone's fear response is different, and it's important to respect their feelings and support them in whatever way they need.

DEALING WITH THE CONFRONTATIONAL

I include this as it is so much more important than most people realize. I have always found that surviving in the wild against nature, the elements, predators, and vectors, is the easier part of any survival. It is surviving the other humans that can be the most difficult, unless you're prepared for it. This is where you develop your leadership spurs. So, here are some thoughts on how to best deal with someone who is hindering or obstructing your efforts. And no, we don't shoot anyone in the face in the civilian world! Be nice and move around them. "Survive with honor" should be your mantra in all things.

Dealing with a disruptive person can be challenging, but it's important to stay calm and focused on your own safety.

Speak calmly and clearly to the disruptive person and try to explain the importance of taking shelter. Let them know that their actions could put everyone in danger and encourage them to cooperate.

If the disruptive person is putting others in danger or refusing to cooperate, seek help from emergency responders or other authority figures. They may be able to intervene and help calm the situation.

Remember, every second counts, so it's important to prioritize your own safety and the safety of those around you. By staying calm, communicating clearly, finding a safe location, and seeking help if necessary, you can increase your chances of survival, even in the presence of a disruptive person.

THE RESET

And then, there is your physicality. I find that sometimes, I have to reprogram or "recalibrate" my mind. For example, using heat, if I find myself feeling too hot, be it fire, which I've survived, or extreme heat, which I've done plenty of, I stop when it makes sense to do so. I close my eyes, breathe in this new reality, telling my mind and body that it is a change, that it will be okay, that I can tolerate more, and I imagine the heat all the way to the extreme of burning my flesh and lungs. Once I feel that extreme potential version of my heat-related situation, then the current heat doesn't seem so bad. And so, I reset the thermometer in my mind-body connection, and I no longer feel the heat as oppressively—because I've faced it worse in my mind.

The simplest way to explain the concept further is the classic special forces planning model:

Number one, go to the worst-case scenario, face it, and make a plan. Then go to the number two scenario of what will likely be the most likely scenario, and make a plan to get things done. It works for me every time.

THE LIVING PHILOSOPHY

A philosophy I have lived with for decades after my first few bouts of combat engagements is pretty simple really, but it gives great confidence and strength in the face of adversity and the unknown. Simply live each day like it's a gift, and like it's your last. Never waste an opportunity to tell folks you love them, to do good, and to help others. It seems silly, but the net result is that it lets you accept that one day, eventually, will be your last. If you live well, you are always ready to meet your maker and, as such, when faced with a challenge, you can rise to the occasion, do what needs to be done, and don't dwell on the what-ifs should something go wrong. So, instead of being distracted by those small things that can worry you, you let go of all fear, and focus moment by moment until the immediate issue is resolved. It doesn't mean you're not afraid, it just means you are present and ready.

SPIRITUALITY

Internally, I have many different mechanisms for dealing with the kinds of stress that come with extreme weather events. I've evolved my techniques to be tailored for each situation in the big picture. I know that there's no one-size-fits-all answer to this question, as everyone's spiritual beliefs and practices are unique. However, here are some general suggestions for a spiritual way to deal with disasters:

1. Pray: Pray for safety and protection, both for yourself and for those in the path of the event.
2. Meditate: Meditate to calm your mind and find inner peace, even in the midst of a chaotic and stressful situation.
3. Practice gratitude: Practice gratitude by focusing on the things that you are thankful for, even in difficult circumstances.
4. Find strength in faith: Find strength in your faith, whatever that may be. Draw on the comfort and support that your beliefs provide to help you cope with the challenges.
5. Connect with others: Connect with others in your community and share your spiritual beliefs and practices. This can help to create a sense of solidarity and support during a difficult time.

6. Practice compassion: Practice compassion by helping those in need, offering support, and showing kindness and empathy to others.

Remember that spirituality can be a powerful tool for coping with difficult situations, including natural disasters. Find what works best for you, and draw on your beliefs and practices to find comfort, strength, and resilience. If all this is a bit too mumbo jumbo for you, I get it. In war, we have a saying: *There are no atheists in foxholes.*

Here are a few short passages from the Bible that may inspire hope and provide comfort during natural disasters:

FROM PSALMS

1. *God is our refuge and strength, an ever-present help in trouble.* —Psalm 46:1
2. *The Lord is near to the brokenhearted and saves the crushed in spirit.*—Psalm 34:18

FROM PROVERBS

1. *Trust in the Lord with all your heart, and do not lean on your own understanding. In all your ways acknowledge him, and he will make straight your paths.*—Proverbs 3:5–6
2. *When calamity comes, the wicked are brought down, but even in death the righteous seek refuge in God.*—Proverbs 14:32

FROM THE NEW TESTAMENT FOUR GOSPELS

1. *Peace I leave with you; my peace I give to you. Not as the world gives do I give to you. Let not your hearts be troubled, neither let them be afraid.*—John 14:27
2. *Come to me, all you who are weary and burdened, and I will give you rest.*—Matthew 11:28

These passages remind us that God can be our refuge and source of strength in times of trouble, and that we can trust in His guidance and protection.

They also remind us to turn to our God for comfort and peace during difficult times and remember, these are not the times for judging others for their beliefs, so long as they are filled with love, kindness, and humanity; they are on the path to goodness, as are all who seek it.

I have faith that our next generation will find a way to survive, no matter what comes. If we stop fighting each other, stop seeing each other as different, but see the world as one planet, one race, and we work for sharing and balance globally of wealth and resources, then we can do it. We can do anything we want if we believe and choose to do so. But it also takes leaders with vision, who can see peaceful ways to bring balance and betterment for all the world, mankind, animal kind, and nature.

NUTS AND BOLTS

First and foremost, as a leader, you have to set the example but you also have to survive to lead. Always plan on *not surviving*. This way you make sure you pass on the knowledge, wisdom, and experience to your loved ones. Knowledge of a secret cache of water dies with you if you die, and that could lead to your family unit perishing, too. Not pushing your kids to know how to do certain things like friction fire, could be their downfall.

Remember, no one should ever be bored, as there is always more to do and learn. However, everyone needs rest, and humans need fun from time to time, so be sure to build those into your plan, as well.

One of the best ways to prepare yourself emotionally for an extreme event is to educate yourself about what to expect and to have a solid plan in place for what you will do. Here are some steps you can take to prepare yourself emotionally:

1. Know the signs: Learn how to recognize the signs for what may be imminent. Understanding what to look for can help you feel more in control and less anxious.

2. Have a plan: Make sure you have a plan in place for what

you will do. This should include identifying the safest place in your home to take shelter and knowing the fastest route to get there. Having a plan can help you feel more prepared and less vulnerable.

3. Stay informed: Keep up-to-date on weather conditions in your area. This can help you feel more in control and less surprised by any sudden changes in the weather.

4. Practice relaxation techniques: If you find yourself feeling anxious or overwhelmed, try practicing relaxation techniques such as deep breathing or meditation. These can help you feel calmer and more centered.

Remember, it's normal to feel anxious or scared when faced with a potentially dangerous situation. By taking these steps to prepare yourself emotionally, you can feel more in control and better able to cope with whatever comes your way.

ACCEPTANCE AND DEFIANCE

No battle plan fully survives first contact with the enemy, so expectations of modifications—or what the military calls Fragmentary Orders (or FRAGOs)—are a staple of special operations missions.

So, expect to have to change your plans if and when disaster strikes, and do not be dismayed or let it deflate your motivation. Whatever you face, meet it head on with all the gusto you can muster, as it may be your last shot at survival, and if that is to be the final engagement, let any who may come across the wreckage, carnage, and devastation see that you gave it your all and fought the good fight until the very end.

Notice I didn't say bitter. To me, that is wrongful thinking. We all will meet our end, and more times than not, it won't be what, when, or how we wanted—that's just the nature of death and dying for most folks most of the time. If you fight your very best, giving it your all, in the face of fearful odds, then there should come a sort of peace in that notion alone. And, for your loved ones, if it is their time, they, too, will know and see in that moment just how much you loved them to fight so viciously for their survival.

I speak on this as I personally feel I'd be remiss in my duties as a teacher to not speak on this very real potential outcome. Part of what gives warriors the strength to persevere and prevail against all odds is that very simple paradox; if you believe, if you live well, if you live ready, at any moment, you can rise up and overcome— or fall with honor and dignity that will outlive you in the hearts, minds, and memories of those who do survive. That's really the most we can hope for. So, make peace with things, so when death comes calling, you are ready for your heaven and, until then, you are ready to give it hell and make him earn that life he seeks to claim. Deny him that, with all your might, letting go of all fear; in that way, and only that way, can you actually have the very best chance to survive.

MENTALIZE IT

To ease the physical discomforts of extremes, such as cold or heat, put your mind to it . . . as the saying goes, mind over matter.

> **Practice deep breathing:** Taking slow, deep breaths can help reduce stress and anxiety which can, in turn, help lower your body temperature.
> **Visualization:** Visualize yourself in a cool (or warm), comfortable place, such as a snow-covered mountain or a cool, shaded forest, or sitting by a vibrant fire. This can help shift your focus away from the heat (or cold) and create a sense of calm.
> **Positive self-talk:** Use positive self-talk to remind yourself that you can handle the heat (or cold) and that it will not last forever.
> **Meditation:** Meditation can help reduce stress and anxiety which can, in turn, help lower your body temperature. Focus on your breath or a calming image to create a sense of relaxation.
> **Listen to calming music:** Listening to calming music can help reduce stress and create a sense of calm.

Take a mental break: If possible, take a break from your current environment and do something that you enjoy, such as reading a book or watching a movie.

MENTAL HEALTH

Everyone has feelings. Emotions are part of being human. Whether you believe in a god or not, we have experienced the same range of emotions. Keeping cool through to the end of the event or your own demise, a level head will extend your life as long as you can keep it. So, mental hygiene is important. Some food for thought, as keeping good mental health during a cataclysmic disaster and mass death can be challenging, but there are some steps that can help. Here are some tips:

1. Acknowledge your feelings: It's important to acknowledge and accept your feelings of grief, anxiety, fear, or anger. These are normal responses to a traumatic event and it's important to give yourself permission to feel them.

2. Connect with others: Connect with loved ones, friends, or support groups to share your feelings and experiences. Social support can be a powerful tool in coping with stress and trauma.

3. Practice self-care: Practice self-care by eating well, getting enough sleep, and engaging in physical activity. This can help to reduce stress and improve your overall well-being.

4. Limit exposure to media: Limit exposure to media coverage of the disaster or mass death. Too much exposure can be overwhelming and increase anxiety and stress.

5. Seek professional help: Seek professional help if you are experiencing severe distress or symptoms of mental illness. Mental health professionals can provide therapy and support to help you cope with the trauma.

For me, it's love. I love my family and the people in my life. When things get bad, I keep my cool to keep them alive. I stay ready to leave this world, so all that remains is to stay focused on your reason for fighting to survive, and that is for them to live.

Now, here's a bunch of other info that doesn't really fit anywhere else, but may be useful for your planning and prep.

The stages of grief, as outlined by psychiatrist Elisabeth Kubler-Ross in her work *On Death and Dying*, are:

1. Denial: In the first stage of grief, a person may feel disbelief or denial that the loss has occurred. They may try to rationalize the situation or block out the reality of the loss.

2. Anger: In the second stage of grief, a person may feel anger or frustration. They may blame themselves or others for the loss, or feel resentful toward those who seem unaffected by the loss.

3. Bargaining: In the third stage of grief, a person may try to negotiate or make deals in an attempt to undo the loss. They may also try to find meaning or purpose in the loss.

4. Depression: In the fourth stage of grief, a person may feel overwhelmed by sadness or depression. They may experience feelings of guilt, regret, or hopelessness.

5. Acceptance: In the final stage of grief, a person comes to accept the reality of the loss and begins to find ways to move forward. They may still experience sadness or grief, but they are better able to cope with their emotions and adjust to life without the person or thing they have lost.

It's important to note that not everyone experiences grief in the same way or in the same order, and some people may not experience all of these stages. Grief is a highly individual process and can take time to work through. It's important to seek support from loved ones or a mental health professional if you are struggling with grief.

First Aid for Weather-Specific Injuries

HEAT

Heat injuries can range from mild to severe, and treatment will depend on the type and severity of the injury. Here are some general guidelines for treating common heat injuries:

1. Heat cramps: Rest in a cool place and drink plenty of fluids, such as water or a sports drink containing electrolytes. Stretching and massaging affected muscles may also help relieve cramps.

2. Heat exhaustion: Move to a cool place and remove excess clothing. Drink plenty of fluids, such as water or a sports drink containing electrolytes. Apply cool, wet cloths to the skin or take a cool shower. Seek medical attention if symptoms do not improve within an hour.

3. Heatstroke: Heatstroke is a medical emergency that requires immediate treatment. Move the person to a cool place and call 911. Remove excess clothing and apply cool, wet cloths to the skin. *Do not* give the person anything to drink. Monitor the person's breathing and vital signs until help arrives.

4. Sunburn: Apply cool, wet cloths to the affected area, and

take over-the-counter pain relievers, such as ibuprofen or aspirin, if necessary. Take cool baths or showers to help soothe the skin and reduce inflammation. Apply aloe vera gel or a moisturizing lotion to the affected area to help soothe the skin and promote healing. Avoid further sun exposure until the burn has healed.

It is important to seek medical attention if you experience severe symptoms, such as blistering, swelling, or fever, or if the burn covers a large area of your body. In addition, it is important to take steps to prevent sunburn, such as wearing protective clothing, using sunscreen, and avoiding prolonged exposure to the sun during peak hours.

SUN ISSUES

Here are some of the best things to have for first aid for your whole family for solar issues. While solar flares and related geomagnetic storms do not directly cause physical harm to people, they can lead to indirect consequences that may require first aid. Power outages, communication disruptions, and general confusion can result in accidents or injuries. Therefore, it is essential to have a well-stocked first aid kit for your family. Here are some items to include:

1. Adhesive bandages: Various sizes of adhesive bandages for covering minor cuts, scrapes, and blisters.
2. Sterile gauze pads: Different sizes of gauze pads for covering and protecting larger wounds.
3. Medical tape: For securing gauze and dressings on wounds.
4. Antiseptic wipes or solution: For cleaning and disinfecting wounds to prevent infection.
5. Tweezers: For removing splinters, ticks, or other foreign objects.
6. Scissors: For cutting gauze, tape, or clothing if necessary.
7. Elastic bandages: For wrapping sprains, strains, or providing support to injured limbs.

8. Instant cold packs: For reducing swelling and pain from sprains, strains, or bruises.

9. Pain relievers: Over-the-counter pain relievers such as ibuprofen, acetaminophen, or aspirin for managing pain and inflammation.

10. Antihistamines: For treating allergic reactions and itching from insect bites or stings.

11. Hydrocortisone cream: For relieving skin irritation, inflammation, and rashes.

12. Burn gel or aloe vera: For soothing and treating minor burns, including sunburns.

13. Oral rehydration salts: For treating dehydration caused by diarrhea or vomiting.

14. Thermometer: For monitoring body temperature in case of fever or illness.

15. Disposable gloves: For protecting your hands when administering first aid.

16. First aid manual: A comprehensive guide to help you handle various medical emergencies.

17. Emergency contact information: A list of emergency phone numbers, including your family doctor, local hospital, and poison control center.

18. Prescription medications: Ensure that you have a sufficient supply of any prescription medications your family members may need.

19. Face masks: In case of respiratory illnesses or air quality issues, have a supply of N95 or similar face masks available.

Remember to periodically check and replenish your first aid kit to ensure that all supplies are up to date and in good condition. Keep your first aid kit in an easily accessible location and make sure all family members know its location and how to use the supplies within it.

BURNS

Not wanting to get too deep into the treatment for burns, this will just serve as an overview to let you know how much you don't know and that you need to study more on burn first aid.

Firstly, most folks know the three degrees of burns:

First-degree burns are simple, red making, skin hurting like sunburns. Treat it with cool compresses of ice or cold water, topical anaesthetic lotions and pain medicine, like aspirin, Motrin, or Tylenol help.

Second-degree burns have blisters and hurt a lot as well. Do not pop the blisters if you can help it. Let the new skin underneath grow and the white blood cells and plasma in the blister are sterile and will protect it. If you need to pop it, do so at the base with a clean needle. Sterilize it with a match or lighter first, let it cool, then pop at the base of the blister near the good skin and let the deflated skin remain as the best bandage you could hope for. If it pops on its own, it does. The treatment for both is a dry, sterile dressing. Wet ones attract dust and dirt and introduce those contaminants into the burn causing infection.

Third-degree burns are when the skin is charred, it doesn't hurt, because the nerves have been burned up. This is a life-threatening emergency.

These burns will let plasma leak out of the body which causes severe dehydration and hypovolemia will ensue and that low blood pressure can cause death. It is important to keep the person covered in dry sterile dressings, get fluids in them the best you can, drinking, enema, or intravenous if you have that ability. And a broad spectrum of antibiotics will be needed; be sure they don't have any allergies to what you intend to give them. They will begin to swell and their skin will be like dried leather or plastic, constricting their circulation.

In special ops medicine, wilderness first aid, and other extreme medicine, you may need to perform an escharotomy, or cutting of the skin to allow the swelling without cutting off circulation. It's beyond the scope of this book to teach these things, but this basic overview should point you in the right direction to learn where you lack and prepare your own pack. That said, let's make sure we double tap the critical things to plan for dealing with burns and wounds.

When preparing an aid bag for treating burns, you should include the following items:

Sterile saline solution or clean water to clean the burn area.
Sterile gauze pads or clean, soft cloths to cover the burn area.
Antibiotic ointment or cream to apply to the burn to prevent infection.
Pain relievers such as acetaminophen or ibuprofen.
Burn ointment or aloe vera gel to soothe the burned area.
Disposable gloves to protect the caregiver from infection.

Burns can range from mild to severe and may cause a variety of injuries such as:

First-degree burns: These are the least severe and affect only the top layer of skin. Symptoms include redness, swelling, and pain. Treatment includes cooling the burn with cold water and applying aloe vera gel or burn ointment.

Second-degree burns: These burns affect the first and second layer of skin and may cause blisters, redness, and pain. Treatment includes cooling the burn with cold water, covering the burn with sterile gauze, and taking pain relievers.

Third-degree burns: These are the most severe and affect all layers of skin and underlying tissues. Symptoms include charring, numbness, and white or blackened skin. Treatment involves seeking immediate medical attention.

Smoke inhalation: Inhaling smoke can cause respiratory distress, coughing, and wheezing. Treatment includes moving the person to a safe area, administering oxygen if available, and seeking immediate medical attention.

It's important to note that severe burns require immediate medical attention, and it is important to call for emergency medical services or seek medical attention as soon as possible.

The best first aid treatments for minor to serious burns are:

Minor burns: For first-degree burns, immediately cool the burned area with cool water for ten to fifteen minutes, or until the pain subsides. After cooling, cover the burned area with a sterile gauze pad or clean, soft cloth. You can also apply a burn ointment or aloe vera gel to soothe the burn. Over-the-counter pain relievers like ibuprofen or acetaminophen can help relieve pain.

Moderate burns: For second-degree burns, which can be more painful and involve blistering, first cool the area with cool water for ten to fifteen minutes, or until the pain subsides. Then, cover the burned area with a sterile gauze pad or clean, soft cloth. Use over-the-counter pain relievers like ibuprofen or acetaminophen for pain relief. If blisters form, avoid breaking them, and do not apply any ointment or cream.

Severe burns: For third-degree burns, which are the most severe, immediately seek medical attention. Do not try to cool the area with water or apply ointments or creams. Instead, cover the burned area with a clean, sterile cloth and go to the nearest hospital or emergency room.

Electrical burns: For electrical burns, make sure the source of the electrical current has been turned off, and seek medical attention immediately. Do not touch the burned area with your bare hands, as the electrical current may still be present.

Remember, burns can be very serious and may require professional medical attention. If you are unsure about the severity of a burn, seek medical attention as soon as possible.

DEHYDRATION

If someone is experiencing extreme dehydration, it is important to seek professional medical help immediately. However, in the meantime, there are a few first aid measures you can take to help alleviate the symptoms:

1. Rehydration solution: Oral rehydration solutions such as Pedialyte or Gatorade can help to replace the electrolytes and fluids lost due to dehydration. It is essential to provide fluids gradually and continuously.
2. Cool compresses: Placing cool compresses on the forehead, neck, and underarms can help to reduce fever and heatstroke symptoms that can accompany severe dehydration.
3. Rest: Encourage the person to rest in a cool, shaded area, and avoid any physical exertion.
4. Monitor vital signs: Monitor the person's vital signs, such as their pulse and blood pressure, and report any significant changes to medical professionals.

If you can, keep an IV set on hand and make sure you're well trained on how to use them. It's high-level training, but it really can save lives.

COLD WEATHER INJURIES AND TREATMENTS

Hypothermia: Hypothermia occurs when the body's temperature drops below normal. The real threat here is that it can be insidious in its onset, making you tired, lethargic, decreasing your mental acuity, causing you to not think clearly with the end result being that it creeps up on you before you know it, and then it's just about too late. Kids, older folks, and those with medical conditions are more susceptible.

Early symptoms: Shivering, fatigue, loss of coordination, confusion, and disorientation.

Late symptoms: No shivering, blue skin, dilated pupils, slowed pulse and breathing, drowsiness, slurred speech, paradoxical undressing, terminal burrowing, and loss of consciousness, often leading to death.

Treatment: Get help *ASAP*. Get to warmth and/or a shelter. Remove wet clothing and replace with dry clothes or blankets, warm drinks (not booze) unless they're unconscious, and gently rewarm the person starting with torso, neck, head, and groin, before going to limbs. Gradually warm the affected area by immersing it in warm water (100 to 105 degrees Fahrenheit) for fifteen to thirty minutes. Do not use hot water or a heating pad, as this can cause burns. Rub gently or try skin-to-skin contact, getting undressed and under the blankets with them to warm them back up. Once rewarmed, keep them that way—they could drop faster and easier on a second round of *cold* exposure.

If no pulse: Consider CPR. This is always true when you witness the cardiac arrest. If you come upon someone in this state, unless you have a lot of resources, the chances of CPR working are slim. But you may do it for your own peace of mind, or for that of others who may be there.

Frostbite: This is an injury caused by extreme cold. It usually happens to exposed extremities such as nose, ears, chin, fingers, and toes. When human cells freeze, they swell just like a can of Coke in the freezer, and then they rupture. Once those cells are destroyed, they usually can't recover. This can lead to the necessity to amputate the dead tissue.

Symptoms: Reduced blood flow to extremities, numbness, tingling, or stinging, aching, bluish or pale waxy skin.

First aid: Get warm, use a dry room if plausible, try warm water if feasible (104 to 106 degrees Fahrenheit). Keep water temp warm, not hot. Try not to walk on frostbitten feet as it can cause more tissue damage. Warm affected body parts by placing them in armpits or groin. Do not rub or massage as that can cause more tissue damage as well.

Trench foot: This is also known as immersion foot. It can happen even in warm temperatures of 60 degrees if the skin is exposed to prolonged periods of wetness, as the body can lose heat twenty-five times faster with wet feet than dry. When this happens, the blood vessels shut down circulation, and tissue begins to die from lack of oxygen, nutrients, and the buildup of toxic byproducts that ensues.

Symptoms: Red skin, numbness, leg cramps, swelling, tingling pain, blisters or ulcers, bleeding under the skin, and gangrene. This is when the tissue dies and begins to become infected, turning the damaged area a dark purple, blue, or even grey.

Treatment: Remove offending wet footgear and socks, dry feet gently, and avoid walking on them until they're healed.

Chilblains: These are caused by the repeated exposure of skin to temperatures just above freezing and as high as 60 degrees. The cold exposure causes damage to the capillary beds (groups of small blood vessels) in the skin. This damage is permanent, and the redness and itching will return with additional exposure. The redness and itching typically occurs on cheeks, ears, fingers, and toes.

Symptoms: Redness, itching, blistering, inflammation, maybe ulceration.

Treatment: Avoid scratching, slowly warm skin, use steroid creams to help with itching and swelling, and keep blisters clean and uncovered or wrap loosely with gauze. Apply aloe vera gel or a mild corticosteroid cream to the affected area to reduce itching and swelling.

It's important to remember that prevention is the best way to avoid cold weather injuries. Dress in warm, waterproof clothing and avoid prolonged exposure to cold temperatures, especially if it's windy or wet.

Let's look at what are some good items to have in your kit for cold weather first aid needs. Cold weather injuries can be serious and require prompt treatment to prevent further damage. Here are some of the best things to have on hand for treating cold weather injuries:

1. Hypothermia wrap: A hypothermia wrap, also known as an emergency blanket, can help to prevent heat loss and hypothermia in cold weather.
2. Heat packs: Heat packs can be used to warm up cold extremities, such as hands and feet.
3. Warm liquids: Warm liquids, such as tea or soup, can help to raise body temperature and prevent hypothermia.
4. Hand and foot warmers: Hand and foot warmers can provide additional warmth to the extremities and help prevent frostbite.
5. Waterproof bandages: Waterproof bandages can be used to cover and protect frostbitten or injured skin.
6. Saline solution: Saline solution can be used to flush out eyes or wounds that have been exposed to the cold.
7. Pain relievers: Over-the-counter pain relievers, such as ibuprofen or acetaminophen, can help to relieve pain and reduce inflammation.
8. Anti-inflammatory cream: Anti-inflammatory cream can be used to treat injuries, such as frostbite or chilblains, and reduce swelling.
9. Cold packs: Cold packs can be used to reduce swelling and inflammation from cold-related injuries.

It's important to note that while these first aid items can help treat cold weather injuries, they do not substitute for professional medical care. If you or someone else is experiencing a cold weather injury, seek medical attention immediately.

STORM-RELATED INJURIES AND TREATMENTS

LIGHTNING

Lightning strikes can cause serious injuries, including burns, cardiac arrest, and neurological damage. If someone has been struck by lightning, it's important to seek medical attention immediately. Here are some first aid steps you can take while waiting for medical help to arrive:

1. Call for emergency medical services: Dial the emergency services number in your area and request an ambulance. It's important to get medical attention as quickly as possible.

2. Check for breathing and pulse: If the person is not breathing or has no pulse, start CPR immediately.

3. Check for burns: Lightning strikes can cause burns, so check for burns and provide first aid as necessary. If clothing is stuck to the skin, don't try to remove it. Cover the affected area with a sterile dressing.

4. Check for shock: Lightning strikes can cause shock, which can lead to fainting or loss of consciousness. If the person is conscious, have them lie down with their feet elevated.

5. Monitor the person's condition: Watch for signs of worsening symptoms, such as difficulty breathing, chest pain, or seizures. If the person's condition deteriorates, be prepared to provide additional first aid until medical help arrives.

It's important to remember that lightning strikes can cause a wide range of injuries, and some of these injuries may not be immediately apparent. If you or someone else has been struck by lightning, it's important to seek medical attention as soon as possible. Even if you feel fine, you may have internal injuries that require treatment. It's often overlooked but if someone has a lightning strike burn, chances are, there is an exit or entry wound somewhere else, so don't forget to look for it. And there is a very good chance, if they survive the strike, they will likely have some internal organ damage.

Here are some of the most common wounds seen in lightning strike patients:

1. Burns: Lightning strikes can cause burns on the skin where the electrical current enters and exits the body. The burns can range from minor to severe and may require medical attention.

2. Cardiac arrest: Lightning strikes can cause the heart to stop beating, leading to cardiac arrest. This is a medical emergency and requires immediate treatment.

3. Neurological damage: Lightning strikes can cause neurological damage, including memory loss, difficulty concentrating, and changes in personality.

4. Muscle pain and spasms: Lightning strikes can cause muscle pain and spasms, which may be temporary or long lasting.

5. Hearing and vision loss: Lightning strikes can cause hearing and vision loss, which may be temporary or permanent.

6. Headache and dizziness: Some lightning strike patients may experience headache and dizziness, which can be symptoms of neurological damage.

7. Gastrointestinal damage: Lightning strikes can cause gastrointestinal damage, including stomach and intestinal perforations.

8. Respiratory damage: Lightning strikes can cause respiratory damage, including lung damage and pneumothorax (collapsed lung).

One of the big indicators that something is going wrong on the inside of a person after a lightning strike is very similar to a burn or crush injury; there is damage to the tissue, and the body is trying to deal with expelling all the dead and damaged cells. This leads to other problems, like organ failure, especially with the kidneys as one of the key signs you'll see if the injured person begins to have dark brown, red, or tea-colored urine. This indicates rhabdomyolysis is occurring, and they need fluids, by mouth if conscious, by IV if you have the training and supplies—but for sure, they need a hospital as they'll likely need high-level care, like dialysis.

Rhabdomyolysis is a medical condition that occurs when muscle tissue breaks down and releases a protein called myoglobin into the bloodstream. Myoglobin can cause damage to the kidneys and other organs, leading to potentially serious complications. This results in a condition which is, technically, the rapid disintegration of striated muscle tissue accompanied by the excretion of myoglobin in the urine. It causes your muscles to break down (disintegrate), which leads to muscle death. When this happens, toxic components of your muscle fibers enter your circulation system and kidneys. This can cause kidney damage, otherwise known as renal failure, which leads to death.

Rhabdomyolysis can be caused by a variety of factors, including:

1. Trauma or crush injuries to muscles
2. Severe burns, like the type of injuries you'd see in a lightning strike
3. Prolonged immobilization
4. Infections, such as influenza or bacterial infections
5. Excessive exercise or overexertion
6. Heat stroke or heat exhaustion
7. Drug and alcohol abuse

Symptoms of rhabdomyolysis may include muscle pain and weakness, dark urine, fever, nausea, and vomiting. In severe cases, rhabdomyolysis can cause kidney failure and other complications, and it requires immediate medical attention.

Treatment for rhabdomyolysis typically involves hospitalization, where fluids are given to flush out the myoglobin from the body and prevent kidney damage. In some cases, dialysis may be necessary to help support kidney function. The reason I dwell on this so much is because being a former special forces medic and a current paramedic, I know this is a most likely result if the person is hit by lightning and they live.

Even if they are up and about at first, chances are it will catch up to them, so the key to survival is knowing this before you see it, and getting after it without delay.

HURRICANES

Since this is not a medical book, I don't dive too deeply here into my medicine. But as a medic, it's in my nature to always address those basics you are likely to encounter within each subgroup of Mother Nature's traumas and drama. The best first aid for wind injuries will depend on the type and severity of the injury. Here are some general guidelines:

Seek shelter: If you or someone else is injured during a wind event, seek shelter as soon as possible to protect against further injury.

Assess the injury: Assess the injury and determine the severity. If the injury is severe, seek medical attention immediately.

Control bleeding: If there is bleeding, control it by applying pressure with a clean cloth or bandage.

Treat wounds: If there are open wounds, clean them with soap and water or saline solution, and cover them with a sterile bandage or dressing.

Treat fractures: If there are fractures, immobilize the affected area with a splint or other suitable object.

Treat hypothermia: If someone is exposed to cold and windy conditions for an extended period, they may develop hypothermia. Remove wet clothing and wrap them in a blanket or warm clothing to raise their body temperature.

Treat shock: If someone is in shock due to an injury, lay them down and elevate their feet to increase blood flow to vital organs.

Remember that wind injuries can be serious, and it's important to seek medical attention if the injury is severe or if there are any concerns about the injury. If you're not sure how to treat the injury, seek advice from a medical professional or trained first aid provider.

TORNADOES

When it comes to first aid, the basics never change: C-ABC, Control Hemorrhage a.k.a.: stop any major bleeding, make sure they have an open airway, ensure they are breathing, plug those air holes in chest if needed. Check again for any massive bleeders, that they are in fact stopped with your dressing(s), and check for all the little ones you may have missed in the big trauma moment. Tornadoes can cause a wide range of injuries, from minor cuts and bruises to life-threatening trauma. To help your mind space for these kinds of injuries, here is a look at the common injuries typically caused by tornadoes:

1. Lacerations and bruises: Flying debris such as broken glass, tree limbs, and building materials can cause cuts and bruises.
2. Fractures and broken bones: People may suffer fractures or broken bones from being struck by falling objects or being tossed around by the strong winds.
3. Head injuries: Traumatic brain injuries can occur if people are hit by flying debris or thrown against hard surfaces.
4. Burns: Tornadoes can cause fires or electrical hazards, leading to burns.
5. Hypothermia: People can develop hypothermia if they are exposed to cold and wet conditions after the tornado.
6. Respiratory problems: Dust and debris stirred up by the tornado can cause respiratory problems, such as coughing, wheezing, and shortness of breath.
7. Psychological trauma: Tornadoes can cause significant psychological trauma, such as post-traumatic stress disorder (PTSD) and anxiety.

It's important to seek medical attention immediately if you or someone else has been injured during a tornado.

Also, expect to see a lot of penetrating injuries resulting from flying debris. In general, penetrating objects such as bullets, knives, or shrapnel should not be removed from the body during trauma care, except in certain circumstances.

If a penetrating object is left in place, it can help control bleeding and prevent further damage to internal organs. Removing the object could cause more harm and increase the risk of bleeding and infection.

However, there are some situations where removing a penetrating object is necessary for the patient's survival. These include:

1. Airway obstruction: If the object is obstructing the airway, it may need to be removed to allow the patient to breathe.
2. Uncontrolled bleeding: If the object is causing uncontrolled bleeding and cannot be controlled through other measures, it may need to be removed.
3. Infection: If the object is contaminated or infected, it may need to be removed to prevent further infection.
4. Object mobility: If the object is moving around, it can cause additional damage and may need to be removed.

In general, if you are providing trauma care to a patient with a penetrating object, it is best to leave the object in place and seek medical attention as soon as possible. A medical professional can evaluate the situation and determine the best course of action for the patient's care.

SANDSTORMS

For sandstorms issues, here are some specific considerations for rendering medical care and what to expect. Sandstorms can cause various injuries, some of which are:

1. Eye injuries: Sand and dust can cause eye irritation and injury, including corneal abrasions, conjunctivitis, and vision loss.

2. Respiratory problems: Inhaling sand and dust can cause respiratory problems such as coughing, shortness of breath, and asthma attacks.
3. Skin injuries: Exposure to sand and sun can cause sunburn and skin irritation.
4. Trauma injuries: Flying debris during a sandstorm can cause head injuries, lacerations, and fractures.

Here are some first aid measures you can render in case of these injuries:

1. Eye injuries: Flush the eyes with clean water to remove any sand or dust particles. Seek medical attention if there is severe pain or vision loss.
2. Respiratory problems: Move to an area with fresh air and rest. Use a mask or bandanna to cover your nose and mouth to avoid inhaling sand and dust. Seek medical attention if symptoms persist.
3. Skin injuries: Apply cool compresses or aloe vera gel to soothe the skin. Drink plenty of water to avoid dehydration.
4. Trauma injuries: Apply pressure to any bleeding wounds and seek medical attention immediately.

POLLUTION

Not a lot different, but there are a few things peculiar to pollution to think about here in regards to medical considerations. The type of first aid measures needed against pollution will depend on the type and severity of pollution exposure. Here are some general first aid measures that may be needed in case of pollution exposure:

1. Move to fresh air: If you are exposed to pollution, move to an area with fresh air immediately to prevent further exposure.
2. Seek medical attention: Seek medical attention if you experience symptoms such as difficulty breathing, chest pain, or severe headaches.

3. Wash affected areas: If you come into contact with pollution, wash affected areas with soap and water to remove any contaminants.

4. Remove contaminated clothing: If clothing is contaminated with pollutants, remove it and wash it separately from other clothes.

5. Use a mask: If air pollution is present, use a mask or respirator to filter out harmful particles.

6. Drink water: If you are exposed to pollution through ingestion or skin contact, drink plenty of water to help flush out toxins.

It is important to seek immediate medical attention if you experience severe symptoms of pollution exposure, such as difficulty breathing or chest pain. In case of an emergency, call your local emergency number. If you're working in and around pollution, pay attention to the OSHA standards. If you live near those places, keep an eye on the CDC and other local entities that monitor and report that info to the public for awareness and safety.

APPENDIX III

Helpful Resources

I f you are interested in being able to completely survive any weather event, the books to have as reference are very much like those needed for complete off-grid living. There are a variety of books that can be useful to educate your family. Here are some types of books that may be helpful:

1. Homesteading and self-sufficiency guides: Books that focus on homesteading and self-sufficiency can provide practical advice on everything from growing your own food to building shelter and managing livestock.

2. Renewable energy guides: If you're planning to live off-grid, you'll need to generate your own power. Books that focus on renewable energy, such as solar, wind, or hydro power, can provide guidance on how to set up and maintain your own power system.

3. Survival guides: Survival guides can provide information on basic wilderness survival skills, such as shelter building, fire starting, and first aid. They can also cover more advanced techniques, such as trapping and hunting.

4. Natural building and design guides: Books on natural building and design can provide guidance on how to build environmentally friendly structures that are well suited for off-grid living. This may include techniques such as straw bale construction, earth bag building, or timber framing.

5. Permaculture and sustainable farming guides: Permaculture

and sustainable farming guides can provide advice on how to grow your own food using sustainable, low-impact techniques. These books may cover topics such as companion planting, soil management, and crop rotation.

Here are some of the best apps to have for every kind of extreme weather event:

1. National Weather Service: The National Weather Service app provides detailed weather forecasts, warnings, and radar maps for all types of weather events across the United States.
2. Red Cross Emergency: The Red Cross Emergency app provides real-time alerts for severe weather, as well as preparedness tips, first aid instructions, and emergency contact information.
3. Weather Underground: Weather Underground provides hyperlocal weather forecasts, radar maps, and severe weather alerts for your specific location.
4. Hurricane Tracker: Hurricane Tracker provides up-to-date information on current and predicted hurricane paths, as well as detailed storm tracking maps and evacuation routes.
5. Dark Sky: Dark Sky provides hyperlocal weather forecasts, precipitation maps, and alerts for your current location.
6. FEMA: The FEMA app provides alerts for natural disasters, as well as tips for emergency preparedness, disaster assistance applications, and disaster recovery resources.

Here are some of the best websites to visit for planning against every kind of extreme weather event:

1. National Weather Service: The National Weather Service provides a wealth of information on all types of weather events, including severe weather alerts, radar maps, and detailed forecasts.
2. Ready.gov: Ready.gov is a website run by the Federal Emergency Management Agency (FEMA) that provides

information on how to prepare for all types of natural disasters, including extreme weather events.

3. NOAA's National Hurricane Center: The National Hurricane Center website provides up-to-date information on current and predicted hurricane paths, as well as preparedness tips and evacuation routes.

4. The Weather Channel: The Weather Channel website provides detailed weather forecasts, radar maps, and severe weather alerts for your specific location.

5. AccuWeather: AccuWeather provides hyperlocal weather forecasts, radar maps, and alerts for all types of weather events.

6. Earthquake Country Alliance: The Earthquake Country Alliance website provides resources and information on how to prepare for earthquakes, including preparedness tips and earthquake safety guides.

Here are a few checklists to review as you make all considerations and preparations that you can; don't forget the stuff that makes life and the world go around.

The best kind of vehicles to own for use in extreme weather depends on the specific weather conditions you may encounter. Here are some examples of **VEHICLES** that are well suited for different types of extreme weather:

1. Snow and ice: Vehicles with all-wheel drive or four-wheel drive, as well as high ground clearance, are ideal for driving in snowy or icy conditions. Examples include SUVs and pickup trucks.

2. Floods and hurricanes: In areas prone to flooding or hurricanes, it's important to have a vehicle with good ground clearance and the ability to navigate through deep water. Examples include trucks or SUVs with snorkel kits and water-resistant seals.

3. Extreme heat: In areas with extreme heat, vehicles with good

air conditioning and cooling systems are important for staying comfortable while driving. Cars with light-colored interiors can also help reflect heat and keep the vehicle cooler.

4. Tornadoes and high winds: Vehicles with a low center of gravity, such as sports cars, are less likely to be affected by strong winds and tornadoes. However, it's important to prioritize safety and seek shelter in a sturdy building if a tornado is approaching.

Remember, regardless of the type of vehicle you own, it's important to maintain it properly and keep emergency supplies in your vehicle in case you encounter extreme weather conditions while driving.

The best kind of **HOMES** to own for living and surviving all kinds of extreme weather depends on the specific weather conditions in your area. Here are some examples of homes that are well suited for different types of extreme weather:

1. Snow and ice: Homes with steeply pitched roofs are ideal for shedding snow and ice, as well as preventing ice damming. In addition, homes with good insulation and energy-efficient windows can help keep energy costs down during cold weather.

2. Floods and hurricanes: Homes that are elevated off the ground or built on stilts are ideal for areas prone to flooding or hurricanes. These types of homes are less likely to be damaged by rising water levels or storm surges.

3. Extreme heat: Homes with good insulation and energy-efficient windows are important for staying cool during hot weather. Homes with light-colored roofs and walls can also help reflect heat and keep the interior cooler.

4. Tornadoes and high winds: Homes with strong, reinforced roofs and walls are important for withstanding high winds and tornadoes. In addition, homes with storm shutters or impact-resistant windows can help protect against flying debris.

5. Wildfires: Homes with fire-resistant roofing and siding,

as well as well-maintained landscaping, are important for protecting against wildfires. In addition, homes with good ventilation and air filtration systems can help protect against smoke and other pollutants.

Remember, regardless of the type of home you own, it's important to maintain it properly and keep emergency supplies on hand in case of extreme weather events.

In addition, it's important to follow local building codes and regulations to ensure that your home is properly constructed to withstand extreme weather conditions.

When it comes to surviving extreme weather events, **SHELTERS** such as caves, silos, and bunkers can provide protection from the elements. Here are some factors to consider when choosing a cave, silo or bunker for survival:

1. Location: The location of the cave, silo, or bunker is important for several reasons. It should be situated in an area that is unlikely to be affected by extreme weather events, such as flooding or landslides. In addition, it should be easily accessible in case of an emergency.
2. Size and capacity: The size and capacity should be sufficient to accommodate your needs and the needs of your family or group. It should have enough space to store food, water, and other supplies, as well as provide adequate living quarters.
3. Ventilation and air quality: Adequate ventilation is necessary to ensure good air quality. This may include ventilation systems or natural airflow.
4. Security: It should be secure and able to withstand potential intruders or threats. This may include security measures such as locks or security systems.
5. Sustainability: It should be sustainable over a long period of time, with access to resources such as water, food, and power. It should also be designed to minimize waste and environmental impact.

Remember, caves, silos, and bunkers are not a perfect solution for surviving extreme weather events, and there are many factors to consider when choosing one as a shelter. It's important to do your research and seek out professional advice to ensure that your cave or bunker is safe and effective for survival.

About the Authors

Jim N. R. Dale is an ex–British Royal Navy meteorologist and founder of British Weather Services. Born in Manchester, England, Jim has held a fascination for the natural environment from his earliest years, with a particular affinity to the weather, for which he feels he has a "sixth sense" with concern to the weather's various elements and how they might interact with us.

During his navy days, Jim was charged with observing, forecasting, and relaying the weather to an assortment of officers and crew in some of the most hazardous parts of the world, including the tempestuous Southern Ocean in and around the Falklands (Malvinas) and South Georgia, at the time of and following the war with Argentina.

In 1987 Jim formed British Weather Services, now the longest established independent meteorological company in Europe. Jim and his company supply a wide range of clients, including television and film production companies, the media, insurers, professional sports clubs, bookmakers, the legal fraternity, retailers, and many other weather-sensitive concerns.

Jim is considered a renowned expert within the UK and beyond, across several aspects of global weather and climate which, in tandem with the knowledge and experience he gained over four decades, led him to write his first book, *Weather or Not? The Personal and Commercial Impacts of Weather and Climate* (LID Publishing, July 2020).

Within two condensed chapters of his book, Jim briefly alluded to some of the contents of this book, recognizing that time was running out in relation to global warming and the consequential perils now facing mankind and planet Earth. It became his desire and mission to not only warn of the consequences, as many others have done, but to impart the knowledge to be as safe as possible and survive the tempests in all their various forms. It was at that point he contacted Mykel to move forward with this important joint venture.

Jim is a family man, married to Kanchanan with three grown sons, Simon, Christopher, and Adam, and a younger daughter, Nanthiya.

Mykel Hawke is a retired special forces officer and senior enlisted with specialties in medicine; his skills include dental, veterinary, obstetrics/gynecology, surgery, anesthesia, tropical medicine, and definitive level trauma care. He maintains his national and state paramedic licenses. Mykel was also a special forces communications specialist with a background covering HF radios up to satellite communications, cryptology, and Morse code. He is currently licensed as an FCC radio operator as well. Mykel was also an intelligence operations specialist, which gave him a steep background in analysis and successful prognostication in matters of unconventional warfare.

Mykel still serves in the reserves, holds a top secret clearance, and spends most of his time supporting and deploying with special operations forces working with active duty special ops from all four branches and PhD engineers to find capabilities gaps and create solutions for them, using his considerable experience in nine different conflicts on every major continent. He is the lead on language translation as a certified project manager, having been rated in seven languages, and keeps up his skills and fitness as a black belt in judo and aikido.

Most people know Mykel Hawke for his works in the media. He has done over sixty television shows, produced and hosted over a dozen, been in a film, and is regularly interviewed by the news media on topics such as survival and special operations. He has published a dozen books on survival, medicine, special ops, and languages; he has also co-authored several, including a thriller series, and contributed to many others. Mykel still does feature media work, such as a recent hit show on Discovery where he was the mentor of four martial artists against three other mentors and twelve other warriors; Hawke's warrior won the competition.

He enjoys fishing from his pier, and has numerous other projects in development from products, films, shows, and his *Survival Homes and Gardens* magazine. Despite all that, he still finds time to help other veterans and support charities benefitting vets, kids, and pets, as those are his passions. But what matters most to Mykel is his family.

Mykel considers himself very blessed to have such a wonderful wife, Ruth England-Hawke. She is not only a well-traveled journalist and seasoned adventurer in her own right, but has worked side by side with Hawke on many dangerous survival missions on television as his co-star. Also, as a father of three incredible sons and three amazing grandchildren, he counts himself very fortunate indeed.

Mykel Hawke is very pleased to partner with the talented Jim N. R. Dale to cover what they both feel is one of the most significant challenges that has ever faced humanity.

We hope this book will serve our fellow mankind as we face unprecedented global climate change. Whether you believe in it being a natural or man-made occurrence, the facts amply show we are experiencing extreme weather with catastrophic impacts . . . and no matter where you are, we all share one thing in common—we all want to survive.